Ivan Pavlov: A Very Short Introduction

VERY SHORT INTRODUCTIONS are for anyone wanting a stimulating and accessible way into a new subject. They are written by experts, and have been translated into more than 45 different languages.

The series began in 1995, and now covers a wide variety of topics in every discipline. The VSI library currently contains over 700 volumes—a Very Short Introduction to everything from Psychology and Philosophy of Science to American History and Relativity—and continues to grow in every subject area.

Very Short Introductions available now:

Available soon:

For more information visit our website

www.oup.com/vsi/

Daniel P. Todes

IVAN PAVLOV

A Very Short Introduction

OXFORD
UNIVERSITY PRESS

OXFORD
UNIVERSITY PRESS

Oxford University Press is a department of the University of Oxford.
It furthers the University's objective of excellence in research, scholarship,
and education by publishing worldwide. Oxford is a registered trade mark of
Oxford University Press in the UK and certain other countries.

Published in the United States of America by Oxford University Press
198 Madison Avenue, New York, NY 10016, United States of America.

Library of Congress Cataloging-in-Publication Data

Names: Todes, Daniel Philip, author.
Title: Ivan Pavlov: a very short introduction / Daniel P Todes.
Description: [New York] : [Oxford University Press] [2022] |
Series: Very short introductions | Includes bibliographical references.
Identifiers: LCCN 2022018906 (print) | LCCN 2022018907 (ebook) |
ISBN 9780190906696 (paperback) | ISBN 9780190906702 (ebook other) |
ISBN 9780190906719 (epub) | ISBN 9780190906726 (ebook)
Subjects: LCSH: Pavlov, Ivan Petrovich, 1849-1936. |
Physiologists—Russia (Federation)—Biography.
Classification: LCC QP26.P35 T6273 2022 (print) | LCC QP26.P35 (ebook) |
DDC 153.1/526—dc23/eng/20220428
LC record available at https://lccn.loc.gov/2022018906
LC ebook record available at https://lccn.loc.gov/2022018907

1 3 5 7 9 8 6 4 2

Printed in the UK by Ashford Colour Press Ltd, Gosport, Hampshire.,
on acid-free paper

For my dear brother Bob

Contents

Contents

List of illustrations

Chapter 1
Winter at Koltushi

New Year's Day 1936 found Ivan Pavlov at his favorite place—his science village outside Leningrad in the countryside at Koltushi. Grieving, exhausted, and excited, he was catching his breath after an extraordinary fifteen months.

They had begun with the Soviet state's effusive official celebration of his eighty-fifth birthday in September 1934. He had then overseen the completion of Koltushi, survived a near-fatal bout with pneumonia in spring 1935, embarked on a triumphant trip to London, returned to Leningrad in early August to star at the spectacularly successful International Congress of Physiology, paid a nostalgic visit to his hometown of Riazan, and, finally, returned to Koltushi for some rest. There, his friend, artist Mikhail Nesterov, painted a portrait of his triumphant host slamming his fists on the table in a characteristic passionate gesture, with his science village, symbol of his historic achievements, in the background.

One month later, in October 1935, his son Vsevolod died suddenly of cancer. Pavlov blamed himself for passing on defective genes and, in his eulogy, proselytized for genetics and eugenics. He then resumed his intense work routine—six days weekly in his three labs, plus the regular Wednesday gatherings with coworkers. In late December, in an annual expression of political incorrectness

1. Mikhail Nesterov, a friend and favorite artist of Pavlov, captured his dynamism and historical stature in his portrait set in Pavlov's apartment at Koltushi.

(he did the same every Easter), he announced a Christmas holiday for his labs and returned to Koltushi. "The damned grippe!" still lingered, Pavlov complained, and "has shaken my certainty of living to one hundred." But he needed at least five more years for his two passions: "to witness the triumph of our [scientific] idea and the fate of my homeland."

His three decades of research on conditional reflexes had reached a critical juncture. Pondering recent experiments on dogs and Koltushi's chimps, Roza and Rafael, he had decided on a bold new turn. Now he gathered his thoughts for April's psychology congress in Madrid, where he would unveil his revised understanding of "the conditional reflex" and innovative lines of investigation on animals' most complex perceptions and behaviors.

As for Russia's fate, he desperately wanted to believe that things were changing for the better. The past two years had been confusing and contradictory. On the one hand, there was the "quiet terror"—the arrest and exile of many "former people" (aristocrats, clerics, merchants, and tsarist bureaucrats), the campaign against religion, and the relentless "reworking of a normal consciousness into a servile one" (Pavlov's words) through the Stalin cult and pervasive propaganda. On the other hand, there was the more moderate Second Five-Year Plan, the end of rationing, the more inclusive Popular Front politics of the Soviet Writers' Congress and Communist International, and, most promising—perhaps, Pavlov suggested, even "the swallows of Spring"—Stalin's sponsorship of a new, more democratic constitution.

In Leningrad, dozens of former people drawn by his reputation as a dissident lay siege to his apartment every day or relayed furtive letters, begging for help. He did what he could, but horrified, moved, and guilt-ridden about his own good fortune, he suffered from insomnia and an irregular heartbeat. Here at Koltushi, he confided to his wife, Serafima, he was "less sensitive to all this," and his heart righted itself.

Pavlov remained at Koltushi that winter longer than usual—for much of January—laboring over two manuscripts. One, entitled "Psychology as a Science," radically revised his long-standing view of the conditional reflex and its place in psychology. The other was his attempt to "do something for religion" in the form of an essay on science, Christianity, and Communism addressed to Vyacheslav Molotov, Stalin's right-hand man.

For the more than six decades since he had defied his father and abandoned the Theological Seminary for studies at St. Petersburg University, Pavlov's passionate practice of science had dominated and integrated his life. He was an outstanding practitioner of the mechanistic vision and new experimental physiology of the mid-late nineteenth century. From the 1890s until 1903 he deployed exquisite experimental and surgical prowess, interpretive ingenuity, and an army of coworkers to analyze the digestive system as a purposive and precise "chemical factory." His masterful *Lectures on the Work of the Main Digestive Glands* (1897) brought him international renown and, in 1904, Russia's first Nobel Prize. That digestive factory, however, was inhabited by a "ghost," the psyche, which continually frustrated (and, for Pavlov, explained) the experimenter's inability to wrest from lab dogs truly "factory-like" results. He then turned to this ghost itself in the three decades of research on conditional reflexes that raised him from a Nobel Prize winner to an international icon.

"Only one thing in life is of essential interest for us," he declared as he began this research—"our psychical experience." His consistent, oft-stated goal was to address "the torments of our consciousness" by fusing the objective and subjective realms—that is, physiological processes and the psychological experiences that accompanied them. These two realms and their relationship constituted what Pavlov termed "higher nervous activity," and by his research—by understanding the psyche in terms of mechanistic, determinist law—he hoped to explain human "torments" and render them accessible to the palliative force of reason.

In his research both on digestion and on higher nervous activity, Pavlov sought precise, consistent, fully determined results in experiments on very complex phenomena in intact animals. A careful experimenter, perceptive observer, and committed truth seeker, he was acutely aware that his results were, in fact, inconsistent. In the digestive research, that awareness had led him to appreciate the great significance of the idiosyncratic psyche. That psyche was itself the explanatory target of his conditional reflexes research—and his attempt to achieve satisfactorily consistent results launched him on investigations of an ever-multiplying set of variables, a constantly expanding range of subjects, including the dynamics and interaction of excitation and inhibition, differing "nervous types" (constitution and

4

temperament) of dogs and humans, the effects of heredity and environment, mental illness, the interaction between individual reflexes and systems of reflexes in the cerebral cortex, and the different ways in which animals make associations between various stimuli.

Pavlov's dogs, salivating on cue to a "Pavlovian bell," entered international scientific and lay culture along with the passionate, charismatic, gray-bearded wizard himself, who was pronounced the Prince of World Physiology to a thunderous ovation at the 1935 International Congress of Physiology, which, in his honor, convened in his homeland.

Yet here lurked a paradox. As Pavlov's closest scientific collaborator Maria Petrova observed, "he was pronounced the greatest physiologist of his time and, nevertheless, in an international setting, remained to some degree alone." Very few of Pavlov's international admirers understood what he was doing, how, and why. That was especially true in the United States, home to the world's largest psychological community, where proponents of the dominant behaviorist orientation adopted Pavlov as one of their own. In so doing, they misinterpreted Pavlov's commitment to objective psychology. For the American behaviorists, "objectivity" signified attention to external behaviors and the exclusion of the "subjective" psyche; but Pavlov was preoccupied with the psyche, and for him objective psychology meant the incorporation of the inner life of animals (including humans) within the laws of higher nervous activity.

This misinterpretation—institutionalized in deeply flawed English-language translations of Pavlov's work and incorporated into common culture in the form of distorted tropes and witticisms—is evident to this day in widespread misunderstanding of Pavlov's practices, goals, and achievements. *Pavlov never trained a dog to salivate to the sound of a bell.* He was completely uninterested in training dogs—except, sometimes, as part of experiments with other goals—and the iconic bell would have proven totally useless to his goals. *Nor did he ever use the terms "conditioned" and "unconditioned" reflex*—which are incorrect, misleading translations of his deeply considered choice of the Russian words *uslovnyi* and *bezuslovnyi*, meaning "conditional" and "unconditional." For Pavlov, the conditional reflex was both a phenomenon and, more important, a *method*—an approach for investigating the psyche. *Nor did Pavlov view his experimental animals as unthinking, emotionless machines.* For him, dogs were simple people, and people were complex dogs. Each had active inner lives that could be understood by reference to the other. Reflecting on experiments and discussing them with coworkers, he constantly interpreted the responses of his lab animals by reference to the life experiences, personalities, feelings, and thoughts of historical and literary figures, his acquaintances, and quite often himself. Powerfully purposeful with a profound moral sense, he struggled to reconcile his belief in free will and personal responsibility with his commitment to determinism.

Pavlov, in short, was not an American behaviorist—his life and thought were deeply embedded in Russian history and culture, especially that of its intelligentsia. Born to a family of priests in provincial Riazan before the serfs were emancipated, he left the Theological Seminary to embrace the new secular faith of science, modernization, and Westernization, made his home and professional success in the booming imperial capital of St. Petersburg, suffered the cataclysmic destruction of his world during the Bolshevik seizure of power and civil war of 1917–21, rebuilt his life in his seventies during the Leninist 1920s, and flourished professionally as never before in 1929–36 during the industrialization, cultural revolution, and terror of Stalin times.

The science of this quintessential objectivist was suffused by the context and common sense of his time and culture and by his own experiences, values, beliefs, and personality. These influences are often recognizable as metaphors in his science. The metaphor of "digestive system as factory" that guided his Nobel Prize–winning studies of the 1890s, for example, was rooted in his response to the industrial revolution of contemporary St. Petersburg—a hopeful response that drew on the scientistic worldview he had imbibed as a teenage seminarian from the "people of the 1860s."

Another central metaphor joined his science directly to his deep love and concern for his homeland: Pavlov's equation of excitation with freedom and inhibition with discipline—and his belief that a balance between the two pairs was necessary for an animal, person, or people to correctly perceive reality and respond appropriately. This metaphor underlay his labs' central line of investigation on nervous types, which both informed and drew on his concern about the "Russian type." Pavlov expressed that concern most explicitly in three speeches during the cataclysmic year of 1918 in which, based on his experiments with dogs, he explained the disasters that had befallen Russia as the consequences of an overexcitable Russian type and held out the hope that his research might enable his countrymen to correct this inborn deficiency. That, indeed, was the central rationale for the founding one decade later of his Institute of Experimental Genetics of Higher Nervous Activity at Koltushi.

Those two final essays composed at Koltushi, then, were Pavlov's last words about two long-standing, interrelated preoccupations. In one, he was fundamentally revising the very concept of the "conditional reflex" to better investigate and understand the psyche in dogs, chimps, and humans, while in the other he was grappling explicitly with the deeper relationship between his science, human psychology (including his own), Christianity, and Bolshevism—that is, with the fate of humanity, and most immediately of his homeland.

The unifying theme in that second essay also expressed an underlying psychological preoccupation that integrated Pavlov's central interests and concerns as a man, a scientist, and a Russian in a single logical–emotional whole. "What," he asks rhetorically, "is the most difficult, really terrible thing in life?" For him, there was only one conceivable answer: *sluchainosti*—happenstance, randomness, accidental events: "*Sluchainosti* of birth—the inheritance of genes and...social class; of environment, initial conditions...of death...illness...of every other misfortune and obstacle in life."

For Pavlov, *sluchainosti* (the singular is *sluchainost'*) were the always-negative, frightening consequences of randomness and unpredictability. As a mechanistic determinist, he did not believe in the ultimate indeterminacy of anything; yet *sluchainosti* plagued a vulnerable individual from outside one's frame of reference, understanding, and control, rendering it impossible to "peacefully and soberly calculate and fulfill my mission in life." That required "a regular, undisturbed course of life and certainty in it. But where can one obtain either?"

The opposite of and antidote to *sluchainosti* was another lifelong keyword—*pravil'nost'* (the adjective is *pravil'nyi*)—regularity, lawfulness, certainty, and correctness—in the organism, in the form of scientific law, and in life. Pavlov demanded these qualities in his daily existence—cherishing order and predictability as a reassuring measure of control over his own explosive nature and the demonstrably cruel randomness of life. *Sluchainost'* was the realm of chaos and vulnerability, *pravil'nost'* that of law, certainty, and control—the province of science. *Pravil'nost'* was also necessary to *tselesoobraznost'*—another lifelong keyword, meaning "purposiveness"—both in physiological processes and in the life well lived.

Pavlov's personality and science were not, of course, reducible to his fear of randomness and drive for certainty. People, and science, are much more complicated than that. This fear, that drive, and those keywords, however, manifested themselves in every dimension of his life: his earliest childhood experiences, his reaction to events, people, and politics; his approach to work, relaxation, and sports; his love letters; and, of course, his science. The keywords *sluchainost'* and *pravil'nost'* worked metaphorically to establish and express intellectual and emotional relations between these various realms. For Pavlov, the threat of randomness and the palliative of lawfulness brought together his reactions to such seemingly disparate subjects as his own emotional explosiveness and bouts of depression, Dostoevsky's novels, the destruction of his university mentor, the framing and interpretation of experiments on digestion and higher nervous activity, the nature of Bolshevism and the plight of his homeland, and the relationship of science and religion.

Pavlov's scientific style, achievements, and goals, then, emerged from a layering of personal temperament and experiences, training and talents, in a particular, Russian setting, during an era in physiology that identified deterministic, mechanistic explanations with "good science," and within a broader social–political culture that joined such explanations to a broader vision and dreams for humanity. He was in this a most fortunate man, a "man of one piece," as one relative put it, whose character, scientific interests, and social–political values developed together in a single, integrated (albeit often contradictory) whole. This reached its apotheosis with conditional reflexes, which was not mere research, but a quest: intense, all-encompassing, alternatingly exhilarating and depressing, in close emotional and intellectual contact with his broader life and preoccupations. The biographical history and intertwining of the emotional, political, and scientific dimensions of this quest shape this introduction to Pavlov's life and science.

Chapter 2
Certainty: religious and scientific

From the day of his birth on September 26, 1849, Ivan Petrovich Pavlov was expected to follow six generations of Pavlov men into service to the Eastern Orthodox Church. His father, Petr Dmitrievich Pavlov, was the much-respected priest of Riazan's Nikolo-Vysokovskaia Church, a pillar of the community who served as spiritual counselor to the local police and regiment and supplemented his emoluments from clerical services with the income from his renowned fruit and berry garden. His mother, Varvara Ivanovna Pavlova, also from a priestly family, was an intelligent, extremely intense woman who, consigned deliberately by her parents to illiteracy, taught herself how to read.

Petr Dmitrievich took pride in the peasant origins and physical strength of generations of Pavlov men, so he worried about the frailty and emotional volatility of his firstborn son. Endowed with a phenomenal memory, Ivan was passionate, willful, and thin-skinned; laughing, screaming, and arguing until the tears fell. He later recalled himself as a "puny, sickly lad" who was constantly at the center of "terrible quarrels," "could spit and swear and fight," and, indiscriminately pugnacious, frequently took a thrashing from bigger boys.

Pavlov's father attributed this worrisome emotionality to Varvara Ivanovna's hereditary influence. After giving birth to her first

2. The Pavlovs in Riazan, probably in 1871. From left: sons Ivan (holding Sergei) and Dmitrii, father Petr Dmitrievich Pavlov, son Petr, mother Varvara Ivanovna Pavlova, son Nikolai or Konstantin. Pavlov's parents may have defaced Sergei's features in this photo after he accidentally killed his brother Petr in a hunting accident.

three children, she had become especially explosive, subject to nervous disturbances and excruciating headaches that often culminated in screaming "frenzies." Pavlov would later describe her lovingly to two biographers, but in ways that led them to characterize her as "abnormally unbalanced." And his father constantly intoned that "there is much of his mother in Ivan!"

Yet the Pavlov line, too, exhibited worrisome peculiarities. Petr Dmitrievich was a steady, disciplined, and unyielding patriarch, but Ivan's two paternal uncles were both defrocked after dramatic displays of what he later termed "internal disorderliness"—which in one uncle took the form of chronic drunkenness and brawling and in the other an apparently uncontrollable passion for strange, irreligious pranks.

Pavlov's lifelong interest in the psyche, then, may well have originated from his experience with his own "morbid, spontaneous paroxysms" (his words) and with the emotional instability of family members. As a student and young professional in the 1870s and 1880s he would several times be diagnosed with "neurosismus" and "hysteria"; as a laboratory chief from the 1890s through the 1930s, his explosive outbursts at coworkers were legendary; and even in his seventies he wondered how such an unbalanced choleric as he could become a successful scientist (a mystery that he finally resolved through conditional reflex experiments on dogs).

The unity of mind and body was part of the Eastern Orthodox culture that Ivan imbibed at home, in his father's church, and for many months at a local monastery. At about age eight, he took a bad fall off a high fence onto a stone platform, breaking some bones, suffering for months from pleurisy and other complications, and becoming so weak that his younger brothers teased him about his frailty. Ivan's godfather, the father superior at nearby Troitsky Monastery, finally took him under wing and subjected him to a strict routine that featured vigorous physical activity during the day and confinement to his room with a reading and writing assignment every night. Upon recovery, Ivan took with him the monk's gift of a favorite book, *Krylov's Fables*, which he kept on his writing table for the rest of his life.

Pavlov left no record about the fear and helplessness he must have felt during the long, uncertain months after the accident, but would speak often about his rescue at the monastery as a turning point in his life. The monk's remedy for the broken boy became a tale and credo about the triumph of self-discipline, work, and regularity over the depredations of *sluchainost'* (chance, uncertainty), a lesson about the unity of mind and body and the benefits of combining physical and mental labor. As a young man, he often honored those lessons in the breach; in later years they became the foundations of his daily existence.

Despite the atheism of his adult years, Pavlov retained powerful nostalgic memories about the spiritual comforts of religious life. He particularly enjoyed reminiscing about Easter. "During the fast, food on the table was scanty (Lenten), the weather gloomy, the church melodies mournful. And then suddenly there began the bright rejoicing of Easter with its clear sunny days, with joyous, cheerful melodies, and with an abundance of tasty delicacies." Asked by a coworker in the 1910s about his attachment to the holiday, he alluded revealingly to an Eastern Orthodox custom that promised children certain protection from ill fortune if they could accomplish one task: "I remember vividly how on Holy Thursday mother would bundle me and my brothers off to church, give us each a candle, and tell us that during the service we must light a candle and then carry it home—and we went and worried that the candle would go out. And these memories are so joyful that I sometimes go to church on Christmas and Easter."

Seminarian in the 1860s

At age fifteen, he entered Riazan Seminary, where the culture featured that familiar emphasis on balance, self-discipline, mind/body integration, and the certainties of faith. Here, some 470 students received a classical ecclesiastical education that emphasized church doctrine and history while also preparing them in history, literature, foreign languages, and math. The natural sciences were represented by one course on physics and cosmography.

Ivan's circle at seminary included his younger brothers Dmitrii and Petr and two sons of the rural clergy, Nikolai Bystrov and Ivan Chel'tsov, who rented corners in the Pavlov home. During their first years there, Ivan, Bystrov, and Chel'tsov ranked at the very top of their class.

Some courses explicitly confronted the heretical views that were becoming increasingly popular within the intelligentsia in the

13

1860s. Moral theology, for example, concerned the history and content of Christian moral doctrine, including "the moral nature of man, his high *dostoinstvo* (moral dignity and obligations) and calling," contrasted the Christian and naturalistic views of virtue, and refuted claims that science contradicted Christian doctrine. Basic theology did battle against materialism, atheism, and pantheism.

Here, too, during the first half of 1867, Ivan received the only formal instruction of his life in logic and psychology. The young cleric Nikolai Glebov explained that human psychology is governed by the complex interplay of body and soul—the former responding passively to external material conditions, the latter acting independent of them. Glebov acknowledged that the digestive, circulatory, and nervous systems all influenced human thought and feelings—even contributing to differences in personality—but emphasized that the soul, a "spiritual, independent, conscious and free force," manifested itself in humans' higher, active self. Stimuli from the external world impress themselves upon our sensory organs, but only the activity of the soul—manifested in attention, representation, and memory—created mental images, ideas, and judgments.

Ivan absorbed Glebov's perspective well enough to earn a grade of excellent in the course, but had by this time embraced a very different worldview under the influence of Russia's "people of the 1860s" (*shestidesiatniki*). Encouraged by Tsar Alexander II's reforms—particularly his emancipation of the serfs, greater latitude for local government, and relaxation of censorship—these thinkers portrayed science as central to the formation of the new type of individual essential for the creation of a new, modernized Russia. Ivan's favorite essayist of the time, Dmitrii Pisarev, argued that these "thinking proletarians" (or "realists"), steeped in the insights, methodologies, and intellectual discipline of science, would sweep away the mystifying muddle of church and state superstition and lead the way to a productive, powerful, and just Russia.

Students at seminary were expressly forbidden to "read books of their own choosing, especially books with ideas contrary to morality and Church doctrine," but in their hunger for precisely such works Ivan and his friends awoke early to join other enthusiasts in front of Riazan's new public library. They eagerly devoured new translations of the "vulgar materialists" Carl Vogt, Jacob Moleschott, and Ludwig Büchner; lectures by French physiologist Claude Bernard (whom Pavlov later called "the original inspiration of my physiological activity"), G. H. Lewes's *Physiology of the Common Life*, Wilhelm Wundt's *Lectures on the Human and Animal Mind*, and Russian physiologist Ivan Sechenov's controversial *Reflexes of the Brain*, in which he argued that even noble thoughts and volitional acts could be explained as reflexive responses.

The worldview that Pavlov absorbed from his circle's discussion of these readings would remain essentially unchanged throughout his life: The scientific method was identical whether one analyzed a rock, a plant, a frog, a human, or human society. True explanations were mechanistic in each case, since organisms (including humans) were complex machines. Science, free of empty philosophizing, was the only true path to plentiful production, social justice, and human progress—that is, to humans' rational control of their own destiny.

His other favorite nonscientific author (alongside Pisarev) was Samuel Smiles, a leading British popularizer of bourgeois Victorian values, whose *Self-Help* and *Lives of the Engineers* emphasized the importance of character, honor, self-discipline, consistency, and purposeful hard work. By imposing their will on nature, Smiles's heroic engineers promoted reliable, gradual social progress. The British author translated into secular language values that Ivan had imbibed from his religious milieu. Indeed,

Smiles's title *Self-Help* was translated into Russian as *Samodeiatel'nost'*—the same term that Glebov employed to denote the active powers of the soul.

For Pavlov, then, science offered a powerful, modern, secular alternative to religion—an all-encompassing worldview that promised certainty in addressing the big questions in life, replaced service to God with service to country and humankind, and provided a moral compass and purpose. The particularities gained nuance as he matured and professionalized, but the religious fervor and all-integrating nature of his faith and commitment to science never changed.

Quitting seminary in fall 1869, he announced that he would instead study science at St. Petersburg University. That decision led to a bitter confrontation with his father—"heated arguments, in which I went too far and that ended in quite serious quarrels." After one year studying for the matriculation exams, he set off with Bystrov and Chel'tsov for the sparkling center of Russian science.

Science in St. Petersburg

"The faculty at this time was in a brilliant state," Pavlov later recalled. "We had a series of professors with enormous scientific authority and outstanding talent as lecturers." The science faculty was indeed dazzling. The lecturer to Pavlov's freshman class on inorganic chemistry was Dmitrii Mendeleev, who had created his periodic table of the elements one year earlier as a teaching aid for this very course. Pavlov's second-year lecturer on organic chemistry would be another eminent scientist, Alexander Butlerov.

Three other important scientists lectured on the biological sciences during his first year. Professor of zoology Karl Kessler, an ichthyologist, had just been elected president of the St. Petersburg

Society of Naturalists and was among Russia's many pre-Darwinian evolutionists. Another was professor of botany Andrei Beketov, who was well on his way to the scientific, organizational, and pedagogical achievements that would earn him acclaim as the Father of Russian Botany. The first-year course on human anatomy was taught by the country's most eminent physiologist, Filipp Ovsiannikov, a member of the elite Academy of Sciences whose research was cited in Lewes's *Physiology of Common Life* and lauded by Pisarev as an outstanding example of Russian scientific achievement.

Yet it was adjunct assistant professor of physiology Il'ia Tsion who became Pavlov's mentor, initiating him into the pleasures of what his protégé termed "the mature mind." The dynamic young professor created "an enormous impression upon all of us [aspiring] physiologists," Pavlov later recalled. "We were simply astounded by his masterful, simple presentation of the most complex physiological questions and his truly artistic ability to perform experiments." Writing to Tsion decades after their work together, he reminisced warmly that "for me, your lectures in the special course at the university and work in your laboratory are among the best memories of my youth."

Only seven years older than Pavlov, Tsion was already renowned for his scientific achievements, having collaborated in the 1860s with both Claude Bernard and Karl Ludwig on studies of the nervous system's role in the self-regulation of physiological processes. In one important contribution, he demonstrated the reflex action by which the depressor branch of the vagus nerve lowered blood pressure by dilating the vessels. In another, he collaborated with Bernard to discover nerves that accelerated cardiac activity.

Trained and highly praised by western Europe's leading physiologists, Tsion was perfectly positioned for appointment to St. Petersburg University's expanding Department of Anatomy and Physiology. When Sechenov resigned his professorship at St. Petersburg's Medical–Surgical Academy—the country's leading medical school and center of scientific medicine (later renamed the Military–Medical Academy)—he acquired that position as well.

Tsion taught Pavlov a "physico-vivisectionist" approach that combined Claude Bernard's basic view of physiology and the organism with perspectives and precision-oriented techniques associated with Karl Ludwig and German physiology. For Bernard and Tsion, the physiologist ignored metaphysical questions about "essences" (there were neither "materialists" nor "idealists" in the lab) and sought to establish precise, consistent, and purposive regularities in organ systems—a level of organization that was "high" enough to capture the vital qualities that defined physiology's basic subject, but "low" enough to establish the determined relations essential to any science. As the science of life, physiology, then, sought the determined laws of such vital processes as circulation, respiration, and digestion. These processes had their own laws, which could neither violate nor be reduced to those of physics and chemistry.

Tsion emphasized the role of methodological developments that were transforming vivisection into an effective tool for experimentation, allowing physiologists to study in isolation the functions of individual organs. (Pavlov's later term for this would be *physiological surgery*.) Using his own two-volume guide to laboratory experimentation, he demonstrated how, by adapting technologies developed in physics—such as Ludwig's kymograph and Etienne-Jules Marey's sphygmograph and cardiograph—physiologists could observe and record experimental phenomena more accurately and even describe their regularities in numerical terms. A talented vivisector, he also taught the ambidextrous Pavlov the surgical skills necessary to this style of physiology.

By January 1874, Pavlov was hooked on research and petitioned the dean for an extra year on fellowship to continue it without the distraction of fourth-year coursework. Granted the extra year (though not the additional fellowship), he used it to collaborate with other students on three original research projects that extended Tsion's investigations of the nervous regulation of organ systems. Two concerned the heart and circulatory system; the third earned the gold medal in a university competition to address the theme (suggested by Tsion) "On the Nerves Governing the Work of the Pancreatic Gland."

Pavlov's plans were now clear and exciting. After graduation, he would serve as Tsion's lab assistant at the Medical–Surgical Academy while earning there the medical degree considered a prerequisite for a professorship in physiology. Soon, however, he was reeling from a most unlikely and traumatic "wild episode," as he later put it. The linchpin of his plans, "this most talented physiologist" Il'ia Tsion, "was chased out of the Academy."

Tsion's shocking demise resulted from the confluence of various factors in a perfect storm: A powerful faction at the Military–Surgical Academy had always opposed his appointment. His provocative, politically conservative public lecture of January 1873 on "The Heart and the Brain"—in which he attacked "nihilists" who falsely viewed physiology as a "materialist science"—alienated much of the faculty and student body. Antisemitism (he was Russia's first Jewish professor) and his personal truculence set many people in all camps against him.

The precipitating event arose from the mundane issue of grades. Although the academy's faculty and curriculum had been reformed in the spirit of "scientific medicine," many students remained unconvinced that "theoretical sciences" such as physiology were directly relevant to medical practice. (Nor, at this time, was there much evidence to the contrary.) Traditionally, then, a student delegation would request that their teacher of a theoretical science forego an exam and give the entire class a passing grade. Tsion, however, refused. The students ignored his warnings and 130 of them failed his final exam in spring 1874.

Encouraged by a bitter attack on Tsion in the populist journal *Fatherland Notes*, student protestors showed up en masse at his first lecture of 1874, heckling him, pelting him with eggs and cucumbers, and demanding his dismissal. The conflict escalated into a general boycott of classes at the academy and then, when security forces forcibly suppressed the protests, spread to Petersburg's other institutions of higher learning, igniting general student discontent and generating more fundamental demands (for example, an end to police surveillance on campus). The authorities finally suppressed the demonstrations, closed the schools, and arrested and expelled thousands of students. Rather than risk further disorder, they insisted that Tsion take some time off. That leave was extended indefinitely and, in fall 1875, he formally resigned from both university and academy.

Pavlov's adored mentor had been humiliated and his career destroyed, and Pavlov's own plans for an extended apprenticeship as Tsion's assistant had evaporated. A few years later, his fiancée noticed that he never spoke about the incident but clearly appreciated her sympathetic sarcasm about "ignoramuses wishing to receive a medical degree for playing billiards." More than forty years later, in a public address on "the Russian mind," he commented bitterly about Russians' attitude toward freedom of speech: "Do we have this freedom? One must say no. I remember my student years. To say anything against the general mood was impossible. You were dragged down and all but labeled a spy."

Crushed emotionally and crippled professionally by this unpredictable turn of events, he headed to medical school in pursuit of his now much less likely goal of a professorship in physiology. He would find other patrons, but never another mentor.

For the next fifteen, very difficult years Pavlov made his own way in the wilderness. He slogged unenthusiastically through medical school (1875–79) while assuming various tasks in labs to make ends meet and continue his research. Through the intervention of a friend, in 1878 he became manager of the small, poorly equipped lab of Sergei Botkin, the academy's eminent professor of medicine, physician to the royal family, and Russia's leading apostle of scientific medicine. With Botkin's support, he entered the academy's Institute of Physicians, where he wrote his doctoral thesis on the centrifugal nerves of the heart (1883), and the following year received a coveted fellowship to spend two years abroad in the labs of distinguished German physiologists Rudolf Heidenhain and Karl Ludwig. He returned to Russia, however, with bleak job prospects and, now married with a son, scratched together a precarious existence as *privatdozent* at the academy, a part-time instructor of medical paraprofessionals, and Botkin's lab manager. In 1888 and again in 1889 he failed in competitions with better-connected candidates for rare professorships in physiology at Tomsk and St. Petersburg universities. Money was so tight that the family was often compelled to live separately. For some months Pavlov suffered from symptoms that convinced him he was dying of tabes (progressive degeneration of the nerves), but was much relieved by a physician's alternative diagnosis of hysteria.

Science, Self, and Dostoevsky

Pavlov expressed his thoughts and feelings during these years about science, human psychology, and his own values and identity in his intimate correspondence with Serafima Karchevskaia, the religiously devout woman whom he married in 1881.

He wooed her initially through a "journal," *Trapped*, in which the centerpiece was an essay, "The Critical Period in the Life of a Rational Person," a clearly autobiographical meditation on his passage from the passions of the youthful enthusiast to the sober pleasures of the professional scientist. The great challenge of the "critical period," Pavlov explained, was preserving the positive features of the "youthful mind" into adulthood. The youthful mind was excitable, passionate, and wide-ranging; engaged in "issues from all possible sciences, philosophical questions about God, the soul, and so forth, about every fact of life." Yet this "lighthearted stroll from one end of the universe to another" also betrayed an ignorance of the great difficulty of attaining a real truth. As the organism aged and youthful excitation diminished, the thinking individual entered a critical period that challenged him or her to find a mature, sustainable form to preserve youthful passion. For Pavlov, that form was systematized, specialized scientific studies, which replaced "the direct authority of sensations" with "conscious, systematized behavior."

Nature, then, "opens the door and reveals the interesting and alluring kingdom of thought," but entrance was granted only to the person who "undertakes serious and difficult work in order to make oneself worthy of it." Such religious allusions permeated Pavlov's letters. He was in part seeking to reassure a deeply religious woman that, despite his atheism, he was reliably moral; yet he was also describing his own struggle to recast in secular form the religious certainties and values of his youth—and the central role of his science in doing so. (His preoccupation with certainty emerged even in one love letter, where he wrote that "my task is to help develop and to protect what exists in your mind and heart from various dangers and random events [*sluchainosti*]." His fiercely independent fiancée objected to this sentiment as patronizing, and it precipitated a quarrel.)

The materialism and scientism of the people of the 1860s would always remain embedded in Pavlov's outlook, but he had abandoned his earlier "youthful" attitude toward knowledge for the "mature" professionalism that he had imbibed from Tsion. He now identified not as an 1860s materialist, but as a scientist—as he would later put it, "a naturalist who investigates life by the method that best leads to the achievement of true knowledge." Yet he had also preserved his youthful passion for the big questions.

The intellectual–emotional issues involved in his identification as a scientist—and their relationship to his interest in the psyche— also emerge in Pavlov's letters to Karchevskaia about Fyodor Dostoevsky, who was at this time intervening forcefully in the pressing ideological, political, and spiritual issues of the day. The young couple discussed the great writer's novels *The Adolescent* and, especially, *The Brothers Karamazov*. Karchevskaia was studying to become a teacher of rural peasants and in 1880 helped arrange Dostoevsky's participation in a fundraiser for needy students, which led to several personal conversations with him that she cherished thereafter as "the most important moment in my religious life." "Our Dostoevsky" (as Pavlov put it) became a point of reference as they grappled with their feelings and beliefs about a set of related, important subjects for them as a couple headed, perhaps, for marriage: faith and religion, reason and science, intimacy and morality.

Reading *The Brothers Karamazov*, Pavlov identified uncomfortably with Ivan Karamazov, whose harsh rationality and inability to make a religious leap of faith condemned him to nihilism and spiritual disintegration. "The more I read, the more uneasy my heart became." Karamazov's "basic nature, or at least his given state, is the same as mine. Obviously, this is a man of the intellect.... The mind alone has overthrown everything, reconstructed everything.... And the person was left wooden-headed, with a terrible coldness in the heart, with the sensation of a strange emptiness in his being." Reason brought Karamazov rich satisfactions—"recall the Great Inquisitor and such great flights of moral thought"—"yet what a [sorry] life" he led. Pavlov described his own plight by citing Karamazov's confession that he would gladly surrender the pleasures of reason for the comforts of faith. Yet, like Karamazov, his nature and life experiences rendered this impossible.

23

He rejected, though, Dostoevsky's view that this problem was rooted in "the triumph of reason." It resided, rather, in "our very nature" and so raised an important challenge beyond the reach of currently "paltry science"—understanding "the human type." Pavlov returned constantly to this same theme. Liberal essayist Konstantin Kavelin had described "the modern Mephistopheles" as the devaluation of the emotional and subjective side of life in the face of recent successes in deterministic natural science. This, he reminded Karchevskaia, was his own "inextricable issue." Yet here again he identified this as a problem with human nature, and so a challenge for science. How did one reconcile a law-governed universe with the special qualities of personal life, in which "these laws are not in effect, where freedom necessarily rules"? Science now offered only "pathetic" explanations, so man remained unable to understand "the sense, the force of his personal aspirations and efforts." Human psychology remained "one of the last secrets of life, the secret of the manner in which nature, developing by strict, unchangeable laws, came in the form of man to be conscious of itself." Returning to this theme a few weeks later, he added, "Where is the science of human life? Not even a trace of it exists. It will, of course, but not soon, not soon."

He also addressed repeatedly Dostoevsky's view that reliable morality was impossible without religious faith. "I myself do not believe in god," he reminded Karchevskaia, but the religious language of his letters both captured his own struggle for secular replacements and reassured her that he shared her basic values and goals. "Unshrinking rationality and a commitment to truth" was the basis of his struggle for personal virtue, for moral dignity. "It is for me a kind of God, before whom I reveal everything, before whom I discard wretched worldly vanity." A commitment and disciplined approach to scientific truth, could, he assured her, provide the same spiritual satisfaction and reliable moral compass as did religion.

Physiological thinking

Pavlov's immersion in a medical milieu for fifteen years contributed much to his developing scientific style. Deepening the sensibilities that he had imbibed from Bernard and Tsion, it encouraged his lifelong emphasis on whole-animal physiology. Surrounded by clinics and practitioners at the country's center of scientific medicine, he constantly bore in mind the relationship of his research to medical practice and approached the experimental animal much as a physician approached a patient. Working on intact animals, he was, like a good physician, constantly aware of the differences between individuals and, as he observed in one early article, of the influence of an animal's "psychic or physiological state" on its responses to experiments.

As manager of Botkin's lab, he was also compelled to develop management skills and think creatively about the organization of scientific work. Incorporation into Botkin's network generated a broad and influential circle of supporters for the young scientist, who, by his fierce loyalty to his exiled mentor, had alienated himself from the city's leading physiologists.

The two years abroad (1884–86) that he divided between Karl Ludwig's lab in Leipzig and Rudolf Heidenhain's in Breslau also proved significant for his thinking about good physiology and a life in science. Pavlov found both scientists inspiring models of the "mature mind," having preserved their childlike enthusiasm, energy, and kindness by placing "their entire life, all its joy and grief, in science." Their scientific practices deepened the perspective Pavlov had imbibed from Tsion. Ludwig emphasized precise measurement and the value of answering physiological questions in quantitative terms and encouraged Pavlov's increasingly self-confident use of mechanistic imagery. Both Ludwig's isolated heart and Heidenhain's isolated stomach exemplified modern animal technologies that permitted experiments on an intact and functioning organ that responded more "normally" than during standard slash-and-stimulate vivisections.

The differences between the two were equally instructive. Ludwig managed a new type of physiological lab that did not yet exist in Russia: a large, well-equipped facility that made efficient use of inexperienced scientists and physicians attracted by the growing prestige of scientific medicine. Appreciating the advantages of its scale, Pavlov was also critical of the distance it created between Ludwig and the lab bench and of his experimental style: "Experiments are generally conducted stingily, the small details of experiments are not especially taken into account." Instead, coworkers used "more or less exact instruments" to quantify experimental results, which Ludwig then "subjected to careful abstract processing in a study."

Heidenhain operated a small workshop lab and practiced a scientific style that Pavlov considered much better for physiology, which required sensitive interpretation of messy data grounded in a grasp of the experimental situation as a whole. Constantly present at the bench and alert to every detail, Heidenhain conducted "experiment after experiment each day," varying its particulars until he "finally made himself the master of the fundamental condition." Only then did he record the results. The superiority of this approach was evident in Heidenhain's repeated correction of Ludwig's conclusions. "We are not physicists," Pavlov observed, "who can extract the numbers from an experiment and then leave in order to calculate the results in an office. The physiological experiment must always depend on a mass of the smallest circumstances and surprises, which must be noticed at the time of the experiment, otherwise our material loses its real sense."

His wilderness years, then, prepared him well to take advantage of good fortune when in 1890–91 another perfect storm suddenly transformed him from a struggling forty-year-old part-time lecturer to director of Russia's best-equipped and -staffed physiological lab. The basic events were these: Botkin assigned Pavlov the apparently thankless task of representing him on the planning committee for a bacteriological institute under the aegis of activist philanthropist Prince Alexander Ol'denburgskii. Hoping to publicize this project and win it state funding, Ol'denburgskii pressed his scientific collaborators for a rapid experimental verification of Robert Koch's new cure for tuberculosis. The tactic backfired (despite hopeful initial signs, tuberculin failed utterly), destroying Ol'denburgskii's credibility within Russia's medical community. This forced the prince to broaden the scope of his institute and seek persons of lesser stature in order to recruit division chiefs. The prince finally prevailed on his cousin, Tsar Alexander III, to finance his new Imperial Institute of Experimental Medicine. After more eminent candidates either refused the posts or were disqualified as non-Russian, Ol'denburgskii offered Pavlov—who had remained loyal throughout this ordeal and enjoyed the support of Botkin's network—the directorship, which Pavlov declined, and then the Physiology Division, which he accepted wholeheartedly.

For Pavlov and his admiring wife, this extraordinary turn was not the result of *sluchainost'*, but rather the long-overdue fruit of his talents and achievements.

Chapter 3
The haunted factory

In his Physiology Division at the Imperial Institute of Experimental Medicine, Pavlov from 1890 to 1903 guided the labors of about 100 coworkers to conduct the investigations of digestion that he synthesized in his *Lectures on the Work of the Main Digestive Glands* (1897), for which he received the Nobel Prize in 1904.

This research was guided by Pavlov's metaphorical view of the digestive system as a "complex chemical factory" (or sometimes, a "laboratory"), to which he ascribed the very same qualities that he did to successful human endeavors: both were "*pravil'nyi*, precise, and purposeful." As he put it in a speech of 1894, "The digestive canal is in its task a complex chemical factory. The raw material passes through a long series of institutions in which it is subjected to certain mechanical and, mainly, chemical processing, and then, through innumerable side streets, it is brought into the depot of the body. Aside from this basic series of institutions, along which the raw material moves, there is a series of lateral chemical manufactories, which prepare certain reagents for the appropriate processing of the raw material."

A series of research questions flowed from this conception: "What is the activity of this factory at full operation, how and by what is it brought into motion, in what manner does one part go into operation after another, in what manner does the work change in dependence upon the type of raw material, does the entire factory always operate with all its parts, or not? ... One cannot doubt that in the investigation of this subject we will find the very same subtlety and adaptiveness of work that strike us in other, better-studied areas of physiology." In laboratory investigations, the factory metaphor expressed and guided the search for *pravil'nyi* results—for precise, repeated (or "stereotypical"), and purposeful patterns in the glandular responses of dogs to varying quantities of different foods.

Pavlov's lab was also a factory of sorts, a site of tightly coordinated activity that he ingeniously organized and managed in order to enjoy the advantages of large-scale production while also making him a constant participant in work at the bench. Combining free informal discussion and authoritarian decision-making, his distinctive managerial style harnessed the "skilled hands" (as he once put it) of a largely untrained workforce to his own scientific vision, multiplying his sensory reach and productivity while ensuring his control over the critical interpretive moments that arose during experiments.

When it opened in 1891, Pavlov's lab occupied five rooms in the single wooden building that housed all the scientific divisions, but by 1893–94 this site could not accommodate the many aspiring coworkers, so he added a new two-story building. This new labor force was created by a state policy to encourage "scientific medicine" by supporting and providing privileges to physicians who spent up to two years in scientific studies and successfully completed a doctoral thesis. These young physicians were largely unlettered in the sciences, so the prospect of defining, researching, writing, and defending a thesis in two years was daunting.

Pavlov's lab system was designed to meet this challenge while making the best use of these untrained coworkers. Taken under wing by an assistant, they were trained in lab techniques, indoctrinated into interpretive models, provided with a thesis topic and a suitable dog for their experiments, constantly informed of ongoing investigations, and encouraged to share their thoughts in an atmosphere of free discussion. If the chief found their experiments especially interesting, he joined them at the bench; and when results proved baffling, the technical or interpretive issues were resolved by the assistant or Pavlov. Reports, articles, and dissertations were composed according to a standardized form and edited (and often in large part written) by Pavlov himself.

Pavlov's presence permeated the lab. Unless he was lecturing at the Military–Medical Academy (the former Medical–Surgical Academy, where he became professor of physiology in 1895), he arrived in the morning, checked the coatrack to ensure that all coworkers were present, and entered the lab to great effect, as longtime coworker Alexander Samoilov described:

> When in the mornings he entered, or, more correctly, ran into the laboratory, there streamed in with him force and energy; the laboratory was literally enlivened, and this heightened businesslike tone and work tempo was maintained until his...departure; but even then, at the door, he would sometimes rapidly deliver instructions regarding what remained to do immediately and how to begin the following day. He brought to the laboratory his entire personality, both his ideas and his moods. He discussed with all his coworkers everything that came into his mind. He loved arguments, he loved arguers and would egg them on.

Those "arguers," however, rarely prevailed. For one thing, their comments frequently elicited an intimidating outburst of sheer fury (often followed by a sheepish apology in front of the entire lab group). For another, although Pavlov appreciated initiative, "he could not give it a wide range," one associate recalled, "since this would interfere with the development of his scientific idea, which proceeded according to a set plan." There were important exceptions, however, when a bold coworker stood up to the chief and changed his mind, or when—as in 1901 to 1903—Pavlov was pondering a fundamentally new turn, recognized his need for special expertise, and recruited a specific coworker for that reason.

Wandering into the lab, one would find coworkers engaged in a seemingly diverse variety of subjects, yet these reflected Pavlov's "set plan"—his standardized approach to the main digestive organs (the gastric glands, the pancreatic gland, and, somewhat less and later, the salivary and intestinal glands). Research on each gland followed a general sequence: demonstrate the exciters and nervous mechanisms controlling it; develop an appropriate operation (or dog technology) to study its functioning in the intact, "normal" animal; establish quantitatively the regular patterns of glandular activity; and verify the stereotypicity of those patterns through experiments on one or more other dogs. Research on the different glands proceeded in parallel, each providing models for research on the others. Alongside these principal lines of investigation, Pavlov often assigned research topics designed to fortify the Physiology Division's institutional position, explore possible new research paths, respond to critics of lab doctrine, or examine puzzling results.

This highly coordinated, factory-style lab system produced many specific knowledge claims—on the specific exciters of the glands, the role of the vagus and sympathetic nerves, the chemical composition of gastric and pancreatic secretions, etc.—while providing Pavlov a panoramic view of the digestive factory at work and supplying the data for his central synthetic claim: that the *pravil'nost'* (regularity, lawfulness) of this factory was expressed in distinctive "characteristic secretory curves" as the glands provided precisely the quality and quantity of secretions for optimal digestion of various foods.

At the center of lab research were surgically modified dogs for use during "chronic experiments." As opposed to "acute experiments"—conducted while the animal was bleeding, writhing in pain, and soon to perish—chronic experiments analyzed ostensibly normal animals over months and often years. Surgical operations to prepare them, then, were developed to satisfy three criteria: the animal must recover to full health and its digestive system to normal functioning, the product of the digestive gland must be rendered constantly accessible for measurement and analysis, and the gland's reagent must be obtainable in pure form, undiluted by food or other glandular secretions.

For Pavlov, as a convinced "nervist," the digestive system could function normally only if these surgical operations left intact the basic nervous relations that controlled physiological processes. Within his factory metaphor, the intelligence of the factory manager was provided—not by the psyche, which played a very different role—but by the exquisitely sensitive nerve endings that sensed the properties of the ingested food and directed glands to produce secretions of precisely the quantity and quality necessary for its optimal digestion.

The simplest procedure was implantation of a fistula—a thin tube that drew a portion of salivary, intestinal, gastric, or pancreatic secretions to the surface of the dog's body for measurement and analysis. Pavlov refined fistulas for each gland to better meet his three criteria. This proved relatively simple with the gastric and salivary fistulas, but the creation of a normal dog with a pancreatic fistula posed great difficulties. Because of the complex "physiological connections of this gland," the pancreatic fistula constantly leaked, causing ulceration and bleeding and undermining the animal's health in dramatic and mysterious ways. Pavlov revised the operation several times, but conceded that this fistula left much to be desired.

A second standard operation was the esophagotomy, which accomplished "the complete anatomical separation of the cavities of the mouth and stomach" by dividing the dog's gullet in the neck

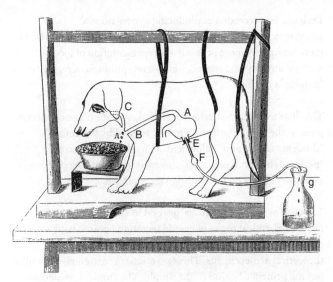

3. When sham feeding an esophagotomized dog with a gastric fistula, the food excites the dog's appetite but, never reaching the stomach (A), falls out the aperture of the esophagotomy (C) and back into the feeding bowl. The resulting gastric secretion ("appetite juice") flows out through a fistula (E) into a receptacle.

and causing its divided ends to heal separately into an angle of the skin incision. Combining the esophagotomy with a gastric fistula allowed experimenters to analyze the reaction of the gastric glands to the very act of eating. Since an esophagotomized dog chewed and swallowed but the food never reached its stomach, this procedure was termed *sham feeding*.

Just before assuming his position at the Imperial Institute of Experimental Medicine, Pavlov had demonstrated that sham feeding an esophagotomized dog excited the gastric glands, allowing him to challenge the dominant mechanistic view that only the food's contact with the esophagus elicited gastric secretion. In the more hygienic facilities of his Physiology

33

Division, he succeeded in producing a more normal esophagotomized animal whose gastric glands responded also to mere teasing. "A more or less lively representation of food," without any physical contact whatsoever, produced a reagent-rich "psychic secretion."

The digestive factory, Pavlov concluded, operated in two different phases: the first, "psychic" secretion, elicited by the animal's "thoughts about" and "enjoyment" of food; and the second, "nervous-chemical" secretion, governed by a food's specific excitatory effects on the nerves in the stomach.

The "Pavlov isolated stomach" devised to study that second, nervous-chemical phase of gastric secretion became both a symbol of his surgical virtuosity and the source of the lab's cardinal theoretical achievements. This operation was extremely difficult, but the principle behind it was simple: The surgeon created an isolated pocket in part of the dog's stomach—and did so in such a way that, after recovery, the entire stomach continued to work normally while this "small stomach" could be studied separately. Food came into direct contact only with the large stomach, but it excited presumably normal gastric secretion in both the large stomach and the isolated sac. Since the isolated sac remained uncontaminated by food and the products of other glands, the experimenter could extract pure glandular secretions through a gastric fistula and analyze the secretory responses to various foods during the normal digestive process.

Pavlov's substantial variation on Heidenhain's isolated sac reflected his nervism. Heidenhain's version involved transection of the vagal nerves, but for Pavlov this rendered it abnormal. Preserving vagal innervation made Pavlov's operation much more difficult—very few physiologists were able to accomplish it—but, unlike the Heidenhain variant, it produced a psychic secretion and secretory responses that varied independently in volume and fermentative strength.

34

Judging the normalcy of these dog technologies would always prove somewhat subjective—an inescapable interpretive moment in Pavlovian experiments that sought "precise, purposive and regular" results from experiments on complex, intact organisms. As Pavlov well understood, uncontrolled variables abounded and identical experiments yielded varying results from dog to dog and day to day.

One well-recognized variable was the animals' response to the operation itself. Pavlov cultivated the image of lab dogs that recovered to lead normal, "happy" lives—but the reality was quite different. The pancreatic fistula would lead to various illnesses and eventually to premature death; the isolated sac would slip, atrophy, or become infected. Pavlov himself once revealed that dogs with an isolated sac tended "to lie on their backs with their legs up," apparently because they experienced "unpleasant or painful sensations when in a normal position." Experimenters, then, necessarily interpreted results in light of the relative normalcy of the animal.

The most important variable was the capricious ghost that inhabited the digestive factory—the psyche. Pavlov himself had demonstrated with experiments on esophagotomized dogs the psyche's importance in gastric secretion. According to lab doctrine, it also influenced pancreatic secretion—either directly (psychic secretion in that gland itself) or indirectly (as the acid from gastric psychic secretion excited the pancreas).

The importance of the psyche was dramatized daily as both motive force and product of Pavlov's lucrative, lab-based business. In 1898 he launched his "small gastric juice factory," which bottled the gastric juice drawn from esophagotomized dogs and sold it as a remedy for dyspepsia. In a small room on the lab's ground floor, five large "factory dogs," selected for their voracious appetite and equipped with an esophagotomy and fistula, stood harnessed to a large table, constantly eating minced meat that then fell through

35

the esophagotomy back into their food bowl. The "appetite juice" that streamed through their fistulas to a receptacle was purified, bottled, and sold as "the natural gastric juice of the dog."

"Appetite is juice," Pavlov liked to say, but his gastric juice factory—this "physiological *perpetuum* mobile," as he proudly put it—captured the duality of the psyche by converting appetite as a psychological force into appetite as a material physiological product. By 1904 (despite an unscrupulous French imitator and the German preference for pig juice), Pavlov's sales of appetite juice had reached 3,000 bottles per year, increasing his lab budget by over 65 percent.

Most important for him, however, was that this natural gastric juice of the dog demonstrated the efficacy of scientific medicine. Physicians had long preached the importance of appetite, and now experimental physiology verified traditional wisdom and improved medical practice by providing appetite juice to patients who, amid the increasing pace and tensions of modern life, paid too little attention to their food, thus depriving themselves of the crucial psychic phase of digestion and so falling ill with dyspepsia.

During chronic experiments that lasted for hours, coworkers constantly encountered this same powerful and idiosyncratic force. As they fed their dog and collected its glandular secretions, the animal's mood, discomfort, and waxing and waning attention, and its response to a movement, a sound, or experimental setting were immediately evident in the volume of secretory flow.

So coworkers were enjoined to study the personality and foibles of their dogs to explain the varying results of identical experiments. These were described in the experimental protocols recorded in the lab notebook for each dog and in coworkers' doctoral theses. "It is taken as a rule in the laboratory to study the tastes of the dogs under investigation," noted one coworker. Some were passionate eaters, others very picky. "Certain dogs are distinguished by a very suspicious or fearful character and only gradually adapt to the laboratory setting and the procedures performed upon them." In doctoral dissertations (which Pavlov edited scrupulously), dogs were described as "peaceful, happy and affectionate," "greedy," "dreamy," "impressionable," "excitable," "stolid," "cunning," and "cold-blooded." Coworker and chief frequently invoked these expressions of "character" or "personality" to explain the absence of factory-like regularity in experimental results.

Only through the "complete exclusion of psychic influence," Pavlov believed, could the otherwise lawful operation of the digestive machine be revealed. So coworkers began conducting chronic experiments in separate, isolated rooms and were enjoined to "carefully avoid everything that could elicit in the dog thoughts about food." Such procedures, however, could not, even in principle, exclude the psyche from experiments. Pavlov and his coworkers were compelled, rather, constantly to correct interpretively for its omnipresent influence.

In his publications, Pavlov portrayed the psyche as a complex actor—regular in its general outlines, but capricious in any individual experiment. On the one hand, it was a constant, objectively measurable participant in the digestive process, producing a predictably ferment-rich initial response to meat and bread (rarely milk, which dogs usually consumed without enthusiasm). One could observe the relationship between the "greediness" with which a dog devoured a meal and the quantity and quality of its secretory response. The "psychic moment," he wrote in *Lectures*, has "acquired a physiological character, that is, it has become compulsory, repeating itself without fail under defined conditions, like any fully investigated physiological phenomenon." From "a purely physiological point of view, one can say that it is a complex reflex." In chronic experiments the psyche "takes form as scientific flesh and blood, transformed from a subjective sensation into a precise laboratory fact." Yet in any specific trial, the activity of this "first and strongest" exciter of secretion depended on the personality, food tastes, and mood of the dog. It was, then, also, a "dangerous" "source of error," threatening constantly to introduce the "arbitrary rule of chance" to experiments and so to produce "completely distorted results."

This contradiction between the psyche as lawful in general but capricious in any specific experiment is apparent in the history of the characteristic secretory curves that provided the rhetorical centerpiece of *Lectures*. These curves demonstrated quantitatively the distinctive course of secretion (with separate curves for volume and strength) elicited from the gastric and pancreatic glands by the ingestion of meat, bread, and milk. For Pavlov, the precision and stereotypical repetition of these different glandular responses to different foods testified to the purposiveness of glandular work. "One must recognize that one or another course of secretion does not exist by chance, but is necessary, useful for the most successful processing of food and the greatest good for the organism."

Yet Pavlov was acutely aware of the psyche's capricious dimension and of his selective choice of experiments to convert into curves in *Lectures*. So, after presenting the impressively similar curves from two pairs of experiments, he added, "Of course, not all experiments are so similar as those given, but if such a similarity is encountered in two experiments out of five, or about that, this cannot, in all justice, but be considered clear proof of the strict lawfulness of glandular work."

The precise meaning of that sentence is clear from the lab careers of the dogs behind these curves, Druzhok and Zhuchka. Pavlov's scientific style featured the intensive analysis of such favored individuals, or (my term) *template dogs*. Temperament, technology, and timing combined to make Druzhok the template dog for the gastric glands, Zhuchka for the pancreas. These dogs' secretory responses became the standard against which those of other dogs were assessed and were enshrined in Pavlov's characteristic secretory curves.

Druzhok (Little Friend) was the first dog to survive the isolated stomach operation and proved a model experimental animal—demonstrating, as Pavlov once put it, "understanding and compliance" during tedious and uncomfortable experiments. The dog's first coworker, Pavel Khizhin, would first tease or feed him, wait about five minutes for the first drops of gastric juice to appear in the special fistula that ran from the isolated sac, and collect secretions at fifteen-minute intervals. This continued for only about two hours in teasing experiments, but when Khizhin fed the dog milk, bread, or meat, both experimenter and animal needed to remain still for five to ten hours to avoid spillage and psychic responses that would skew the data. The compliant Druzhok lay peacefully throughout these trials, even sleeping for five to seven hours at a stretch.

Khizhin's attempts to distinguish between the secretory effect of psychic and nervous-chemical processes were complicated by Druzhok's "unusual impressionability and broad self-esteem." After a few teasing experiments, Khizhin noted, the dog understood that he would not be fed and "turns his snout away." So, no psychic secretion. As Pavlov explained, "The dog is an intelligent animal and is angered by this ruse no less quickly than a person would be." When Khizhin sought to avoid a psychic response by feeding Druzhok stealthily through a cannula into the isolated stomach, the dog often responded to the furtive movements with a greater secretion than when the food was ingested normally.

Despite these difficulties, Khizhin produced in 1894 a much-praised dissertation that enshrined Druzhok's responses in the lab's first characteristic secretory curves. In collaboration with Pavlov, he established that the curves for the strength of secretion were distinctively different for milk, meat, and bread. But those for amount of secretion fell into two different groups—"psychic" and "usual"—that were distinguished not by the type of food ingested, but rather by the animal's psychological response. One could not be certain, even during the second phase of digestion (when the nerves of the stomach presumably responded only to the specific exciters in different foods) that "the psychic excitation of the animal" had been entirely excluded.

This condemned Druzhok to further operations. The lab's prize dog now was esophagotomized and its large stomach fitted with a gastric fistula. The animal's next coworker, Ivan Lobasov, could now acquire a purely psychic secretion (by feeding through the mouth) or, presumably, a purely nervous-chemical response, by feeding through the new fistula. Pavlov assigned him the task of bringing more "consistency" to the psyche's digestive role, incorporating it more fully into the lawful digestive factory. Relying on trials during which Druzhok was fed while asleep, Lobasov and Pavlov concluded that, by incorporating a standard psychic response to the different foods, they could now reinterpret Khizhin's results for both amount and fermentative power as responses to meat, bread, and milk. (Druzhok responded most voraciously to bread, followed by meat; experiments on milk were terminated after three trials because, after exposure to the nonfat variety, he refused to eat it.)

By this time (1895–96), Lobasov noticed that leakage from the isolated sac "ate away terribly at the skin of the belly…and disturbed the dog terribly, threatening its health." (That introduced an important uncontrolled variable into the trials.) When Andrei Volkovich inherited Druzhok in 1897, the dog's secretory responses had become too erratic for lab service.

Volkovich began studying a second dog with an isolated sac, Sultan, allowing Pavlov to write in *Lectures*, "Lately, another dog with an isolated sac…is stereotypically reproducing all the main facts collected on our first [such] dog." Results with Sultan and

Druzhok certainly bore some resemblance to each other, but they differed sufficiently for chief and coworker to refrain from comparing the two in the usual, rhetorically effective form of secretory curves. Druzhok remained the anonymous template dog behind all the curves for gastric secretion in *Lectures*.

In 1896, Pavlov assigned a favorite coworker, Anton Val'ter, to do for the pancreas what Khizhin and Lobasov had done for the gastric glands. Like Druzhok, Val'ter's dog, Zhuchka, proved an ideal lab animal. "It is in the steadfastness of this dog," Val'ter wrote, "that one must find the essential reason for the great regularity of the results obtained upon it." Although Val'ter had not completed his thesis when Pavlov's *Lectures* went to press, his results with Zhuchka already delighted the chief, who used them to display and explicate in *Lectures* the characteristic curves for pancreatic secretion.

Val'ter's and Pavlov's interpretation of experimental trials with Zhuchka led them to the same basic conclusion as had Lobasov's with Druzhok: "Under identical conditions," the pancreatic gland responded differently and with stereotypical "precision and lawfulness" to the ingestion of different foods. This resulted from a combination of psychic secretion and the specific excitability of the gastrointestinal tract.

Also like Khizhin and Lobasov, Val'ter's construction of stereotypical secretory curves relied on interpretation of uncontrolled variables. These included the water content of the animal's body, various factors that influenced the gastric secretion that excited the pancreas, and especially the psyche. "This psychic exciter," Val'ter conceded, "is, by the very nature of things, difficult to subordinate to the control of the experimenter." The animal's "interest" in food varied during and between experiments, inevitably influencing both gastric and pancreatic secretion. With Pavlov's help, he accounted interpretively for the differing experimental results, finding within them the contours of a precise and purposive mechanism. Zhuchka became the Druzhok of the pancreatic gland.

Only after the publication of *Lectures* did Pavlov assign two coworkers to verify Val'ter's findings on other dogs. The disappointingly variant results were attributed in each case to these dogs' physiological particularities and "individual personalities." In a passage doubtless coauthored by the chief, one coworker explained that although dogs lacking Zhuchka's fortunate qualities could not express the same *"pravil'nost'* in the work of the gland," Val'ter's analysis "preserves its correctness even today."

This, then, was the experimental reality behind Pavlov's cryptic comment in *Lectures* about his curves on the factory-like operation of the glands. His phrase "two out of five" referred (precisely) to his choice of two of Khizhin's five experiments on the amount of gastric secretion elicited by a meal of meat. Khizhin had attributed the varying results of these experiments to Druzhok's differing psychic responses; Pavlov chose the two in which those responses were almost identical. The "about that" referred to his choice of two experiments (or, one might argue, of six experiments) of Val'ter's twenty-four. "The powerful impression of such an almost physical precision in a complex organic process," Pavlov enthused, "is one of the pleasant compensations for sitting many hours in front of the glands at work."

We can imagine Pavlov as he writes this passage—searching through his coworkers' data, finding just the right two pairs of experiments, and reflecting about what he had done. He was engaging in a certain sleight of hand, choosing experiments that made his case most convincingly. He signaled this to his readers, I think, both because it seemed the honest thing to do and because he was comfortable with his interpretation of experimental data. He was, after all, following Bernard's dictum to present one's "most perfect experiment as a type"—that is, to choose the experiment that had been most effectively stripped of the "numberless factors" concealing the determinism of physiological processes. Pavlov was no doubt confident that, were he to show all his results to an open-minded and experienced physiologist and have the opportunity to explain the complexity of chronic experiments, the variations in personality and mood from dog to dog and day to day, and the other influences that were inevitably at play in experiments on intact, complex organisms, such a physiologist would accept his choice of model experiments. (And, indeed, although his raw data did not compel belief in a precise and purposive digestive factory, they did cluster in a manner that allowed one, if so inclined, to discern the contours of a factory concealed within.)

Pavlov apparently did just that when, in 1901, he used ten dogs over ten days to present a live version of his *Lectures* to physiologists Robert Tigerstedt and Johan Johansson, representing the Nobel Prize Committee. He did not, apparently, demonstrate the existence of precise, factory-like secretory curves, but he did convince his visitors that, as they reported, "gastric secretion has a different course for different substances" and that the psyche played an important role in glandular secretion. "We have thereby received a new intimation," Tigerstedt wrote, "of the close dependence in which mind and body stand in relation to one another."

This was the first of four years in which Pavlov was nominated for the Nobel Prize, which he finally received in 1904. One weakness of his candidacy was discomfort with his unique, factory system of laboratory production, which gave pause to Nobel committee members guided by an image of the heroic lone investigator. Pavlov himself had written relatively few works, and his *Lectures* was based almost entirely on the research of his coworkers—as he proudly proclaimed, it was "the deed of the entire laboratory." Was this book, then, truly his own, or was it, as one committee member apparently objected, a mere "compilation of the experimental dissertations upon which they are based?" The argument of Pavlov's strongest proponent, Robert Tigerstedt, ultimately carried the day: Pavlov had "awakened" these researchers' interest in the subject; designed and conducted the operations on experimental animals; and conceived, organized, and closely supervised laboratory experiments and lines of investigations. His coworkers' research was entirely permeated by his "guiding idea." Although he alone would never have been able to generate the colossal material produced by the lab, the techniques and knowledge produced "must therefore, to a substantial degree, be seen to constitute one single man's intellectual property."

Only as he was completing *Lectures* in 1897 did Pavlov address in detail the relatively minor salivary glands. He and his coworkers discovered that, unlike gastric and pancreatic secretion, the nervous-chemical and psychic phases of salivary secretion were essentially identical, differing only in amount. Whether placed in the dog's mouth (in a physiological experiment) or waved provocatively in front of it (in a psychic experiment), "everything edible elicits...a saliva rich in mucin, and everything rejected [e.g., acid or sand] yields a watery saliva."

Pavlov initially agreed with coworker Sigizmund Vul'fson that the psyche here was rendering a "judgment," that here "psychology almost entirely overshadows physiology." The research and perspectives of two coworkers recruited for their special expertise, however, complicated the picture considerably. Anton Snarskii argued that, from a modern (Wundtian) psychological perspective, psychic salivation was a low-level process that represented not a conscious judgment, but rather "the automatic reproduction of an established [physiological] reflex; an act completely devoid of any element of conscious choice." Ivan Tolochinov demonstrated that the psychic secretion to food obeyed the same strict regularities as did the knee and nictating eye reflexes familiar to psychiatrists. In each case, if unreinforced by the underlying stimulus (feeding or an actual hammer blow to the knee), the psychic response grew gradually weaker with repetition and finally vanished; and it was easily renewed by application of the underlying stimulus. Pavlov had been more reluctant than some coworkers to view psychic secretion as simply a reflex, but was impressed by these and other experimental regularities that suggested a deterministic, physiological process at work.

He was unconvinced, however, by Tolochinov's view that the essential difference between the physiological and psychic reflex was that the latter acted "at a distance." For Pavlov, rather, the fundamental distinction was this: the physiological reflex was an unvarying "unconditional" response to the essential qualities of the substance—to the qualities related to the physiological role of saliva; while the psychic reflex was a "conditional" response to an otherwise neutral stimulus that had become associated with those qualities—for example, to a color, shape, or smell; a noise; or the sight of the person who usually fed the animal. Reporting in French to the Northern Conference of Physiologists in 1902, Tolochinov introduced the chief's definition of *conditional reflex* (*réflex conditionnel*).

For Pavlov, both the promise and the peril of research on psychic secretion resided in the conditionality—the "inconstancy" and "apparent capriciousness"—of the relationship between stimulus and response. Did this conditionality reflect an inherent indeterminacy in the psyche, or did it represent the animal's complex but determined adaptation to the subtlest change in its conditions, to changing signals about available food or an approaching predator? As he put the question in his first international talk on the subject in Madrid (1903), "Is it possible to bring together all these ostensibly chaotic relationships into some kind of framework, to make the phenomena reproducible, and to discover their rules and their mechanism?"

He was sufficiently confident in an affirmative answer to include "study of questions of experimental psychology in animals" in that year's report on lab activities. The ghost in the digestive machine now became the explanatory target for more than three decades of intense and exhilarating research.

Chapter 4
Pavlov's quest

For more than thirty years, from 1903 until his death in early 1936, Pavlov pursued his quest to understand the mysteries of the human psyche—to constrain it within deterministic, mechanistic law. *Psychic secretion* now became *conditional reflex* (CR), but it also *remained* psychic secretion. For Pavlov, the CR and the idiosyncratic psyche were two dimensions of the same phenomenon, for which he adopted the term *higher nervous activity*. He sought not the replacement of one with the other, but rather their fusion (*slitie*)—the fusion of the objective and subjective, the physiological and psychological.

"If one gives them their psychological term," he announced as he launched this research, "the phenomena of the conditional reflex are precisely...associations." For Pavlov, the CR was both a phenomenon and, even more important, a *method* for analyzing the underlying mechanisms of personality, learning, expectation, emotions, and all the qualities that he had long associated with psychic secretion and that many psychologists attributed to associations. Over the next decades, he and his coworkers would pay even greater attention to the psychological qualities of their dogs than they had during research on digestion. They would discover that dogs were heroic and cowardly, crafty and dull, diligent and lazy, freedom loving and passive, mentally robust, fragile and diseased.

Pavlov enunciated this goal clearly and consistently. He informed the International Congress of Physicians in Madrid in 1903 that he was determined "sooner or later" to "bring the objective results" of CR experiments "to our own subjective world as well, and this will at once and brightly illuminate our mysterious nature, elucidating the mechanism and the real-life meaning of that which occupies the human mind increasingly more persistently—man's consciousness, the torments of his consciousness." In his Nobel Prize speech one year later, he repeated his intention to bring the "strictly objective methods" of natural science to bear upon the one thing that "interests us in life"—"our own psychic content." Speaking to London's scientists and physicians in 1906, he expressed confidence that the "objective material" of CR experiments would provide the basis of psychological knowledge and so "a significant part of the real answer to those torturous questions which have occupied and tormented human beings since time immemorial."

Two decades into his investigations, writing after the carnage of the World War and Russia's civil war, he introduced his *Twenty Years' Experience of the Objective Study of Higher Nervous Activity* (1923), with these same sentiments: "Man,...driven by some dark forces acting within him, causes himself incalculable material losses and indescribable sufferings through the horrors of wars and revolutions that send him back to bestiality. Only the last science, the exact science about man himself—and the most reliable approach to it is via the road of the almighty natural science—will guide him out of the current darkness and cleanse him of all the present indignity in the sphere of interhuman relations."

To follow Pavlov's research trajectory, we must constantly bear in mind the nature of his "objective" data (the saliva drops generated by thousands of CR experiments) and his goal (an explanation of the "mysteries of the psyche"). The path he envisioned to get from one to the other—or, rather, to accomplish their "fusion"—can be broken down into three basic steps: First, discover the (presumably fully determined) regularities in the salivation elicited by thousands of experiments on the formation, variation, and extinction of CRs. Second, use these regularities (patterns) to build a model of the unseen processes in the higher nervous system that produced them. Third, and finally, use that model to explain the behavior, affect, psychological experiences, and personality of experimental animals and, ultimately, humans. In practice, these three steps proceeded in parallel over the course of three decades, with each set of conclusions constantly being revised and compelling changes in the other two. As in his digestive research, but on a qualitatively greater scale in the 1920s and 1930s, Pavlov pursued his grand vision through the coordinated labor of an army of coworkers in specially equipped labs.

In his talks and articles of the early 1900s, Pavlov discussed a number of step 1 regularities: If the experimenter exposed the dog to a previously neutral stimulus that acted on its sense organs— the beat of a metronome, a flashing light, electrical shock—and fed the dog or exposed it to acid at the same time, this created a conditional stimulus (CS) that produced the same reflexive response (a salivary secretion) as did the feeding or acid (the unconditional stimulus, or US) itself. That CS was weakened (elicited fewer saliva drops) if used at the same time as another neutral stimulus. A *strong stimulus* such as a buzzer became a CS more quickly and elicited greater salivation than a *weak stimulus* such as a flashing light. (This became the *law of strength*.) If one CS elicited, say, five drops of saliva and a second CS elicited four drops, the two would (in principle) elicit nine drops of saliva if deployed simultaneously (the *law of summation*). If a metronome (CS) was repeatedly sounded without presentation of food (the US), its beat eventually ceased to elicit salivation. It had become not merely neutral, but a conditional inhibitor (CI)—so if it was now sounded simultaneously with an active CS, the combination elicited less salivation than that CS alone.

On closer inspection, even these relatively simple regularities revealed complexities and variations—and so questions requiring interpretation and further research. Why did two dogs respond to the same CS with differing amounts of saliva? Why did it take fifteen repetitions of the metronome without reinforcement for this CS to become extinct in one dog, but forty-five in another? For Pavlov, every drop of saliva must have a mechanistic explanation, so each variation provided a starting point for new experiments, for the discovery of additional laws. The differences between dogs' performance in identical experiments, to take an especially important example, was addressed systematically in the 1920s and 1930s by investigations of "nervous types."

These studies of the CR as a phenomenon shaded seamlessly into the use of the CR as a method for step 2: constructing a model of the unseen nervous processes that underlay experimental results. What, exactly, transpired in the higher nervous system when a CR was formed and extinguished? What were the precise mechanisms and properties of the dog's "analyzers"? (This was Pavlov's term for the unified nervous mechanism, beginning with the sensory apparatus and ending in the brain, through which the animal received information about the external world.) Pondering such questions, Pavlov developed models to explain the salivary patterns generated by increasingly complex experiments.

One important example that originated in the first decade of CR research was what Pavlov termed *differentiation* (the physiological dimension of the psychologist's "discrimination"). The basic experiment was simple: The experimenter sets a metronome beating at sixty strokes per minute (M60) and feeds the dog. After a number of repetitions, the hungry animal salivates consistently to that sound. The metronome has become a CS, and a CR has been formed. During each trial, the experimenter measures the amount of salivation to gauge its strength. She (these trials were conducted largely by the first generation of women coworkers in Pavlov's lab) now slows the metronome to forty beats per minute

(M40) and does not feed the dog. Initially, the dog salivates to the sound of the slower beat, but after a number of repetitions, it ceases to do so. M60 has remained a CS, but M40 has become a CI. In the years before World War I, the lab used such experiments to determine the acuity of the dog's sense of time and its ability to distinguish among colors, distances, temperatures, shapes, and different points on its body.

Differentiation experiments also contributed to the lab's developing model of higher nervous processes. Why, just after M60 was established as a CS, did the beating of the metronome at *any* speed initially elicit salivation? And why, after a series of repetitions during which only M60 was reinforced with feeding, did the dog cease to salivate in response to any speed other than M60? Pavlov concluded that any stimulus initially "irradiates" across the entire cerebral cortex—leading to general excitation— and only then, in a second phase, "concentrates" at one particular point. He also reasoned that this process of differentiation resulted from "the struggle and collision" of what he viewed as coequal nervous processes—excitation and inhibition. Thus, the beating of the reinforced M60 excited the entire cerebral cortex, but after a series of trials in which other speeds were not reinforced by feeding, these other speeds became CIs and generated an inhibitory wave that suppressed excitation at all speeds other than M60. The ability of the dog to differentiate between M60 and M40, then, depended (according to lab doctrine by the 1910s) on the collision and struggle of excitatory and inhibitory processes. That model changed in 1922, when a particularly talented and determined coworker, Dmitrii Fursikov, convinced Pavlov that discordant data—which had been underplayed and explained away to save the collision and struggle model—supported a more dialectical view of "mutual induction" between excitation and inhibition.

The final, climatic step 3 in Pavlov's agenda was to integrate that model of nervous processes with the behavior, affect, and

4. A dog stands in an isolated chamber in the Tower of Silence. The controls at left enabled experimenters to expose animals to various stimuli without entering the chamber.

psychological experiences of his experimental animals, and ultimately of humans—that is, to fuse the two dimensions of higher nervous activity. How did he seek to traverse the great interpretive distance from step 2 to "the torments of [human] consciousness"—to achieve (as he put it confidently in 1932) "a merger of the psychological and the physiological, of the subjective and the objective"? This question takes us to the heart of his scientific style and the meaning for him of the term *objective*.

For Pavlov, the "objectivity" of his approach resided in two of its characteristics: it proceeded on the basis of a concrete, measurable, physiological phenomenon (salivation), and he moved interpretively *from* this objective phenomenon *to* the subjective realm. This he opposed to *subjective* psychological methodologies that proceeded from postulates about the animal's "internal, subjective world." His notion of objectivity, then, most emphatically did not mean—contrary to a common misconception and to the American behaviorism of his day—that he doubted the existence of the subjective psyche, discounted its importance, believed that scientists should focus only on external behaviors, or himself proposed to ignore the subjective realm. He put it bluntly at a meeting with coworkers: "It would be stupid to reject the subjective world. It clearly exists, of course. Psychology, as a formulation of the phenomena of our subjective world, is an entirely legitimate thing and it would be blind to quarrel with it.... The question is how to analyze this subjective world."

In the early years of his research, Pavlov famously banished subjectivist formulations from the lab: "We strictly forbade ourselves (we even had fines imposed in our laboratory) to use such psychological expressions as 'the dog figured it out,' 'it now intends,' 'it now desires,' and so on. Finally, all the phenomena we were interested in appeared before us in a different light." He replaced psychological terms with physiological ones—for example, *teasing* became *excitation at a distance* and *pain* became *destructive irritation*. In that same spirit, he referred in articles and talks to "so-called psychic" and "so-called *dushevnye*" (literally "of the soul") phenomena; and by the early 1920s he banished those words from his lexicon as well, replacing them with two well-considered terms: First, *higher nervous activity*, which served as an umbrella for both physiological processes (excitation, inhibition, irradiation, concentration, and so forth) and the so-called psychic (or so-called *dushevnye*) that constituted their subjective dimension. Second, beginning in 1923, in a bit of outreach to the American behaviorists whom he considered his most promising audience among psychologists, he equated higher nervous activity with *behavior*. Yet he defined that latter word quite differently than did the behaviorists: for Pavlov, behavior encompassed both the organism's external and its internal (including subjective) responses to its environment.

The banishing of subjective terminology proved brief. Once he had established a physiological lexicon and model of higher nervous processes through which he could approach the dog's affect and behavior objectively, subjective language and anthropomorphic descriptions of lab dogs became central to his research—indeed, these constituted his principal explanatory target.

This was most evident in informal conversations with coworkers and in comments to unintimidating audiences such as the Society of Russian Physicians and the general public, particularly after exciting experimental results achieved with coworkers Maria Erofeeva and Maria Bezbokaia in 1911–12 bolstered Pavlov's confidence that he was on the road to a grand physiology of mind and emotion. In Erofeeva's experiments, after an electrical shock was repeatedly paired with feeding, the hungry animal eventually responded to the shock with salivation rather than an avoidance reaction. Even when ratcheted up to painful levels, shock, the experimenters concluded, had become a CS. This, Pavlov enthused in informal conversations (including one with visiting British physiologist Charles Sherrington), demonstrated how the Christian martyrs "endured terrible torture, and did so smiling." Just as Erofeeva's dog was so hungry that it responded to electrical shock paired with food as a CS, so did the Christians smile while enduring terrible torture because their "enormous moral excitation inhibited, eliminated, the feeling of pain." (In professional presentations, his interpretations were much more restrained.) In Bezbokaia's research, the experimenters used an electrical rod and the dog's learned hostility toward Pavlov to keep the animal in a state of rage and study how that anger influenced its CRs. For Pavlov, this "experimental investigation of emotions," by clearly revealing the dog's heightened salivary responses when behaving aggressively, explained why "a person consumed by passion, for example, jealousy, blames his unhappiness on an innocent person or even on an inanimate object."

As his research developed, Pavlov became increasingly confident and explicit in these interpretive habits—fusing physiological and psychological processes through analogy and moving easily between explanations of dogs and humans. This is especially striking in his articles "The Purpose Reflex" (1916) and "The Freedom Reflex" (1917), in three public addresses of 1918 in which he used lab experiments to analyze Russia's catastrophic plight; publications about nervous types and psychiatry in the 1920s and 1930s; his lab notebooks, recorded meetings with coworkers, and informal comments throughout the 1920s and 1930s.

Pavlov, like Darwin, believed firmly in the continuity of the human psyche with that of higher organisms, and he constantly related the behavior and affect of his experimental dogs to that of his human acquaintances and to himself. Coworker Max Gubergritz noticed in 1915 that, during the experiments that led Pavlov to his concept of the *freedom reflex*, the chief frequently compared the lab dogs "with characters from literature, especially from the works of his favorite writers." Addressing *The Law-Governedness of Mental Life and the Connection between the Scientific Laboratory and Life* (1931), Pavlov used the travails of a lab dog to analyze the mental breakdown of his old friend Bystrov at St. Petersburg University. Just as the dog's "unbearable torment" when confronted with experimental tasks too complex for its nervous constitution elicited "barking and howling as if it were being cruelly tortured," so Bystrov's inability to master the complex principles and procedures of university science studies resulted in "the difficult, tormented state of his hemispheres during their work." In both dog and man, this resulted in "melancholy with excitation."

His favorite subject for such anthropomorphic reflection was himself. During experiments, he frequently commented about similarities between the animals' responses and his own sensations and experiences. "That which I see in dogs," he explained to one reporter, "I immediately transfer to myself, since, you know, the basics are identical." This was both a way to understand the experiments themselves—of reasoning from observable patterns of salivation to the unseen higher nervous processes that lay behind them and the subjective experiences that these produced—and an effort to understand himself scientifically. Such canine-inspired musings included reflections about his nervous type, his style of thought, the influence of aging on the relative strength of his excitatory and inhibitory processes (and, so, on his thinking), his self-experimentation on free will, and his memories of his favorite, deceased son.

The lab's experimental protocols incorporated this same anthropomorphic perspective. As research took off and Pavlov standardized experimental note taking, coworkers recorded not only the time of the trial, the nature of the exciter used, and the amount of salivation generated, but also, in a final column, "other observations." These routinely included such information as the dog "ate greedily," "reacted defensively" (or "aggressively"), "exhibited negativism," or "sighed." By the early 1920s dogs were characterized according to their nervous type and described using the full panoply of human characteristics. In one 1933 article for Pavlov's closely edited house journal, a leading coworker described his dog as an opportunistic sycophant who "knows how to serve."

Pondering the lab dog Pingel in the early 1930s, Pavlov jotted in his notebook, "A Napoleonic type. When free [that is, when not strapped to the experimental stand]—extremely mobile and greedy. In the stand—very peaceful, almost motionless, a small and inconstant secretory reaction to conditional stimuli; a positive movement reaction (to food) is almost absent. Approaches the food trough haltingly and in a demonstratively very slow manner. Then eats greedily, and licks its chops for a very long time, even licking its feet." The goal of his grand quest was to understand the higher nervous processes that produced this greedy, aggressive, and crafty Napoleonic creature—and, by extension, Napoleon himself.

Pavlov's holy grail, then, was the conceptual integration of physiological and psychological phenomena, which he attempted to accomplish by identifying patterns in the formation, variation, and extinction of CRs; by developing a conceptual model of the higher nervous processes that generated those patterns; and by using that model to explain behaviors, personalities, and various subjective states. The fundamental criterion of success remained the same as in his digestive research: the ability to encompass, contain, and so explain experimental phenomena by a limited number of basic principles.

In her memoirs, Petrova captured perfectly the logical consistency in the evolution from the early Pavlov, who banished subjective terminology, to the later Pavlov, who employed such terminology unapologetically: "He ceased to fear psychological terms, since he felt the strength to call them to battle, to confront a new reality, and through this confrontation to discover the path of further work. The goal of the entire enterprise revealed itself at the moment when victory was assured and the unity of the system became obvious to all."

As for the ontological relationship between physiological and psychological processes, Pavlov took a reserved, positivist position. In the early 1900s, he briefly participated in a study group that discussed philosophical theories about the mind/body relationship, but soon quit, preferring to "study the factual aspect" of that question. He did, however, continue to ponder this question privately, jotting in his lab notebook around 1912 that "we consider all so-called psychic activity to be a function of the brain mass, of a defined mechanism, that is, of an object conceived spatially. But how can one place upon this mechanism an activity that is conceived psychologically, that is, nonspatially?" A believer in free will and personal responsibility (and himself extraordinarily goal directed), Pavlov also continually reflected about how to define such freedom and reconcile it with mechanistic determinism and the dynamics of CRs.

In professional settings and publications, he adopted the positivist stance that he (like Bernard and Tsion) considered appropriate for a scientist. When pressed by his audience at Petrograd's Philosophical Society in 1916 about how, precisely, knowledge of CRs would explain psychological experiences, Pavlov acknowledged that "the relationship between nervous phenomena and the subjective world is quite complex" and "to move from one to the other is very difficult." Conceding that he knew little about the various theories, he insisted that this was unimportant. "I have always limited myself only to practice, methods. I cannot agree that my approach represents pure materialism. I am only pursuing an approach useful for investigation."

Pavlov maintained a positivist faith that the accumulation of facts over time would itself eventually illuminate the precise relationship between nervous processes and the subjective realm. In the meantime, in practice, he looked for parallels between the nervous processes transpiring (according to his model) during CR experiments, on the one hand, and the behavior, affect, and personality of experimental animals on the other—and explained the latter as an expression of the former. Cowardice was the result of chronic overinhibition, negativism was the psychological expression of the *ultraparadoxical phase* in higher nervous activity (during which a CS produced an inhibitory response and a CI produced an excitatory response), and so forth.

Pavlov's quest set him on an exciting and fruitful three-decade journey to the horizon. He discovered much of interest along the way—facts, methodologies, relationships, and insights of enduring value—but the ultimate destination continually receded behind an endless landscape of new and perplexing complexities. He was frequently excited and heartened by experiments that produced satisfyingly consistent results and surprising discoveries, the development of new and powerful explanatory principles, the revelation of new research perspectives, and indications that his research might have clinical value (for example, in the analysis

58

and treatment of mental illness). Yet he was also frequently disheartened, even depressed, as the much-trumpeted results and interpretations of earlier experiments were overturned and apparently solid ground dissolved, as the experimental data (and so the psyche) failed to conform to even increasingly elaborate mechanistic models.

Pavlov always responded to these doubts with new experiments—and a single encouraging result sufficed to restore his enthusiasm and confidence. The journey was long, but surely he was on the right path! But therein resided the paradox of his quest: in true positivist spirit, he was convinced that the more experiments he and his coworkers conducted, the more facts they collected, and the more patterns they were therefore able to establish, the closer he would come to his goal. Experimental data would reveal a set of basic laws and a model of higher nervous activity that would, in turn, explain the behavior, personality, and affect of his dogs and, eventually, humans. As he put it in a discussion with coworkers about the vexing question of the dynamics of excitation and inhibition, "We elicit various conditions under which the excitatory and inhibitory processes manifest themselves. Everything comes down to this. The time will come—and it will be such a wonderful moment—when suddenly everything becomes clear, when we will know precisely all the conditions that elicit the process and weaken it."

Yet just the opposite proved true. Over the years, as his lab enterprise expanded and the number of coworkers, dogs, and experiments swelled, the resultant avalanche of data overwhelmed and defeated his every attempt at systematization. Experimental results simply would not fit snugly into even the most imaginative framework. More experiments, in other words, just compounded the problem.

Pavlov's interpretive response to these confounding junctures was always to enlarge the explanatory frame by introducing a new variable that might resolve discrepancies. Here he followed the lead of the eminent French physiologist Claude Bernard: determinism presumably reigned, so if the same experiment produced varied results, there must be some uncontrolled variable. Pavlov thought in broad biological terms about CRs and the nervous system, viewing them as means through which organisms maintained balance both as an integrated whole and in relationship to a changing environment. Constantly imagining the dog in its experimental stand as an animal (or human) in nature and society, he regularly introduced new variables from the broader world beyond the lab. Thus enriched, his explanations continually expanded in scope at the cost of precision and refutability.

This dynamic defined the underlying logic of Pavlov's investigations as they unfolded over three decades, imparting power, paradox, and pathos to his quest. The five "basic principles" through which he attempted unsuccessfully to contain his experimental data when he first tried to write a monograph on CRs in 1917–21 expanded constantly over subsequent years. For example, the fact that different dogs required differing numbers of trials to form and extinguish CRs—combined with their observably different personalities—gave rise in the 1920s to the doctrine of nervous types. Delineation of the various nervous types would, presumably, explain the varied responses of different dogs to the same experiment. Yet experiments designed to establish a typology continually revealed more discrepancies and more variables, so the number of differing nervous types grew from three to more than twenty-five. And still the data were not contained, nor did the results of CR experiments map neatly onto his animals' personalities. His attempt to encompass contradictory data through new variables drawn from a broader biological and social perspective generated new basic principles and lines of investigation on the relationship between excitation and inhibition, the phases of hypnosis and sleep, the interplay of nature and nurture, mental illness in dogs and humans, and the analytical and synthetic qualities of the cortex. Each expansion of the interpretive frame produced valuable observations and insights, each revealed new perspectives that excited the chief and urged him on, and each failed to bring him and his companions any closer to the horizon.

Pavlov's visit to the West in 1923 whetted the appetite of his foreign admirers, who pressed him for a synthetic monograph. Having tried and failed to write one earlier, he had instead, as a "temporary solution," published his talks and articles in *Twenty Years' Experience of the Objective Study of Higher Nervous Activity* (1923). Now he tried again—but again found his manuscript unsatisfactory, discarding page after page and launching major revisions. He finally completed the book only under duress. Gleb Anrep, a former coworker now serving as a reader in physiology at University College, London, delivered a series of lectures about CRs to audiences in England and the United States. In a clearly orchestrated effort, Pavlov's closest Western colleagues informed him excitedly about the great interest in those lectures and suggested that, in the absence of a monograph by Pavlov himself, Anrep might write one. Would Pavlov, they asked, object to that? Thus motivated, he accepted funding from London's Royal Society and began sending manuscript chapters to Anrep, who served as translator. The Russian and English editions both appeared in 1927. (In order not to undercut readership for this volume, publication of the English edition of *Twenty Years* was postponed until 1928.)

Composed reluctantly at a time when uncertainties were mounting and his ideas were in flux, Pavlov's *Lectures on the Work of the Cerebral Hemispheres of the Brain* lacked the elegance and powerful argumentation of his earlier *Lectures on the Work of the Main Digestive Glands*. This volume, rather, captured the state of his research perfectly, offering a compelling general vision and a wide range of intriguing experimental findings, but no synthesis of the two. The first two lectures explained his basic vision and approach, lectures three through ten presented the basic picture of higher nervous processes as Pavlov had understood them in 1921, and lectures eleven through eighteen were devoted to the contradictions, complexities, and new lines of investigation that had emerged since. The book ended not in triumph, but in confession: "Earlier, of necessity we artificially oversimplified, schematized the subject. Now, having some knowledge of its general foundations, we are surrounded—nay, crushed—by a mass of details demanding explanation."

That same besieged, confessional note pervaded Pavlov's remarks to a meeting of his coworkers in December 1926 intended to celebrate the impending publication of his book: "I must thank you for all your work, for the mass of collected facts—for having superbly subdued this beast of doubt. And now...I hope this beast will retreat from me."

Periodically acknowledging his obligation to write another, genuinely synthetic monograph, he never made a serious effort to do so. The temporary solution represented by *Twenty Years* thus became permanent, and successive editions added reports and articles that recorded the experimental and conceptual developments, and the constantly emerging conundrums, of his later research—but without any pretense of systematizing them.

Pavlov's enthusiasm for his quest remained undiminished to the end. During that last winter at Koltushi, he was enlarging the frame again: awaiting the results of a breeding experiment designed to clarify the roles of heredity and environment and even revising his definition of the CR in light of his experiments on chimps and engagement with the insights of Gestalt psychology. Surely, he mused just before his death, if he could live just another five years, he would witness the decisive triumph of his "scientific mission."

Chapter 5
Come the Bolsheviks

During Imperial Russia's last decades, Pavlov lived his vision of the good life. Most important was his science, the center of which was his large, bountifully funded and staffed Physiology Division at the Imperial Institute of Experimental Medicine (IEM). When he became professor of physiology at the Military-Medical Academy in 1895, he acquired also Tsion's former lab there. His Nobel Prize in 1904 soon brought election to Russia's elite Academy of Sciences and another small lab.

The good life was purposeful, precise, and regular. Constantly moving with his swift stride according to a strict schedule, unswervingly committed to his science, Pavlov was not only a charismatic and effective researcher and lab manager, but also the embodiment of energetic optimism and efficient industrial culture. His dynamic purposefulness was manifest in the daily and annual routines that he lived like clockwork—speeding by foot between venues, entering the academy's lecture hall at precisely nine o'clock and his IEM lab just as the cannon at the Peter and Paul Fortress announced the noon hour. Dinners began at six o'clock sharp and the players of *durachki* (a card game) who arrived Sunday evenings at his apartment knocked on his door at precisely nine o'clock.

His annual schedule was equally *pravil'nyi*. In early June every year he boarded a train for his *dacha* (country home) in Sillomiagi (Estonia), where he remained until late August. Here he adopted an unvarying, precise routine featuring gardening, swimming, and hotly competitive games of *gorodki* (Pavlov was a master of this traditional Russian sport that involves throwing a heavy bat at configurations of large wooden pins). Science was off-limits. Pavlov read literature and philosophy, discussed art and music, and socialized with other members of the cultural elite (mostly artists) during months designed to compensate for the imbalances of urban professional life and restore harmony to mind and body.

Pavlov described himself as "a Russian liberal," but his support for gradual evolution to a constitutional monarchy put him to the right of the Constitutional Democrats (Kadets), the liberal party popular with his colleagues. With rare exceptions, he had neither time nor inclination for political activities. Stung by Russia's defeat in the Russo-Japanese War (1904), he earned a black mark with the tsarist secret police during the revolution of 1905 for helping to organize an illegal union of professors. Shortly thereafter he ran for the Duma as a candidate of the center-right Octobrist Party, receiving the most votes of that party's slate in his overwhelmingly Kadet district, but failing to win a seat. Soon disillusioned with politics, he again, as one friend put it, "shut himself off" in his scientific research."

His political ideology featured scientism and a fervent patriotism, with a deep romantic attachment to Russian life and nature (as portrayed by the art he collected) and identification with the strength and international stature of the Russian state. Yet his patriotism coexisted with an uncomfortable conviction that, probably for historical reasons, the "Russian type" was inferior to the English and German. "When the negative features of the Russian character—laziness, lack of enterprise, and even a slovenly approach to every vital work—provoke melancholy moods," he wrote in 1916, "I say to myself, No, these are not our real qualities, they are only the damning inheritance of slavery."

World War I aroused Pavlov's patriotism. He enthusiastically endorsed Russia's war aims and, with two sons at the front, followed the news avidly. As defeat turned to debacle, he scorned Tsar Nicholas II as an "idiot" and "degenerate." Yet he scolded his liberal colleagues in the Kadet Party when they attempted to recruit him. "Don't you understand that you are committing a crime," his wife recalled him saying, "arranging a revolution during wartime! This will lead to no good! No, I will never participate in the ruin of my homeland!" The revolution of February 1917 and abdication of the tsar left Pavlov extremely pessimistic, yet he warmed to the social democratic Provisional Government's promise of expanded freedoms and support for science. Continued social dissolution and military defeats, however, soon confirmed his deepest fears.

Revolution and civil war, 1917–21

For Pavlov at age sixty-eight, the Bolshevik seizure of power in October 1917 and the long, bloody civil war that followed was a catastrophe that utterly destroyed his world. "He talked constantly about the death of our homeland," recalled one close friend, and "regarded the Bolsheviks with hostility and distrust." That hostility was fanned by growing anarchy and material privation, and by the Bolsheviks' nationalization of scientific institutions, granting of independence to parts of Imperial Russia's empire, and agreement to what Pavlov considered a dishonorable separate peace agreement with Germany. At the graveside of his best friend, artist Nikolai Dubovskoi—who died of a heart attack shortly after the Reds occupied his hometown—he grieved for Russia as well: "I envy you. You no longer witness…the ever-increasing destruction and disgrace of our homeland."

Pavlov's circle was annihilated and his family deeply scarred. The Pavlovs' youngest son, Vsevolod, who had interrupted his studies toward a diplomatic career to serve as an officer at the front, informed his parents of the "anarchistic hurricane" devouring the army and joined the White resistance. Their favorite son, Viktor, who combined his father's passion for science with his mother's religious faith, died of typhus on his way to join the Whites. Pavlov's brother Sergei, a priest in Riazan, was arrested by the Reds in 1919 and worked to death in a labor camp. The Bolshevik political police (Cheka) searched the Pavlov home several times and briefly detained both Pavlov and his eldest son, Vladimir.

The Civil War produced disastrous living conditions in the hungry capital (renamed Petrograd during the war), and the life of scientists grew extremely grim. The Bolshevik state, its survival hanging in the balance, wasted few resources on a scientific community that it considered, with good reason, politically hostile. From his spacious quarters at the Academy of Sciences' residence, Pavlov witnessed the disastrous state of affairs for Russia's scholarly elite. One-third of his fellow academicians died. Two perished of cold and hunger in apartments just above and below his own. Others watched helplessly as the Communist authorities housed strangers in their quarters.

Pavlov scavenged for firewood and fed his family from a garden he tended at the IEM. The Bolsheviks confiscated his Nobel Prize money. By 1918, lab work was grinding to a halt. Assistants were in short supply and the dogs starved. (As they did so, they served for experiments on the effect of starvation on conditional reflexes.) "Work has almost completely ceased," Pavlov lamented, and the dark, cold Petrograd winter was approaching. "There are no candles, no kerosene, and electricity is provided for only a limited number of hours. Bad, very bad. When will there be a turn for the better?"

In early spring 1918 he expressed his anguish in a poem in prose:

Where are you, freedom, the eternal siren of human beings, from [those] of beast-like nature to the most complete exemplar of the human spirit? Where are you, genuine, authentic [freedom]; when will you come and remain with us always? Alas! We are doomed to await you at the end of your long and continuous struggle with your implacable foe, the bridle—your struggle in the family, in society, in the state, in all of humankind and in our very soul....

You will arrive, pacified and wonderful...only at the very end of this struggle...when you and your rival extend to each other the hand of peace, embrace in friendship, and, finally, in kinship, as two halves, you merge into a single whole. And this moment will be the beginning of the highest human culture, of the highest human happiness.

In three public lectures that year he made clear that the central metaphor here underlay and united his basic view of the nervous system, animal and human psychology, and the good life; and he expressed the close relationship between his research on conditional reflexes and his concern with Russia's fate. For Pavlov, "authentic freedom" was possible only when a person or people correctly perceived reality and acted accordingly—and this, in turn, depended on a lawful and correct balance between "excitation, or freedom in the broad sense" and "inhibition, or discipline, the bridle." Drawing on his lab experiments, he explained Russia's national tragedy as the result of Russians' chronic imbalance toward excitation, which rendered them incapable of realistic responses to life's demands.

This analysis rested on his experiments and a concept that was increasingly attracting his attention: nervous types. Dogs varied greatly, for example, in their ability to differentiate between two different speeds of a metronome. The necessary "order, measure, and timeliness" to correctly differentiate between the two—and, in general, to accurately perceive and meet the challenges of the external world—could only be achieved by a balanced interaction between excitation–freedom and inhibition–discipline. The absence of inhibition and predominance of excitation produced internal "chaos," an "uncultured type lacking correspondence with reality."

There existed two basic nervous types, Pavlov explained: The even-tempered type, on the one hand, possessing a good balance between excitation and inhibition, performed well in differentiation experiments and proved itself "more perfect and adapted" in life. The excitable type, on the other hand, performed poorly in such trials. "Insufficiently cultured," it "reacts to external phenomena nonsensically." (The lack of symmetry in his reasoning is revealing: at this juncture in Russian history, he was clearly unconcerned with an excessively "inhibited type.")

This experimental truth for dogs applied equally to individual humans and peoples, as was clear from the differences between the English and Germans—whose cultures, laws, and scientific achievements attested to balance—and the Russians, with their overly indulgent childrearing, erratically enforced laws, and preference for windy philosophizing over the sober truths of specialists.

This chronic problem had taken especially dramatic form in the "tragic, calamitous events" of "our revolutionary time." For Pavlov, the reckless pursuit of the February and October revolutions despite the obvious threat of an implacable wartime enemy, Trotsky's absurd negotiating stance during the Brest–Litovsk negotiations with Germany, and the dissolution of the empire all testified to the unrealistic responses of an unbalanced people.

The most telling example was Bolshevism itself. While European social democrats rightly sought to protect industrial workers from the abuses of capital, their Russian counterparts pushed this to a grotesque and unrealistic extreme: "We have overextended this idea into the dictatorship of the proletariat. We have placed the brain, the head, below and the feet above. That which constitutes the culture, the intellectual strength of the nation, has been devalued, and that which for now remains a crude force, replaceable by a machine, has been moved to the forefront. All this, of course, is doomed to destruction as a blind rejection of reality." Russia's only hope resided in science. Invoking experiments with Petrova's lab dog Gryzun, Pavlov assured his audience that "after a certain amount of practice, training," the inhibitory process could be improved substantially. "Despite all that has occurred, we should not lose hope."

He ended his final lecture by reciting his poem, which expressed his vision of the happiness that awaited humanity when excitation–freedom and inhibition–discipline were finally integrated and reconciled. "But gentlemen!" he concluded. "I am devoured by torturous doubt. Are such a merger and such happiness possible for the Russian and the Slav in general, or is it impossible?"

When the tide of Russia's civil war turned decisively toward the Red Army in spring 1920, Pavlov sent the Soviet of Peoples' Commissars (Sovnarkom) a pained letter requesting that the state "permit" him to explore the possibility of emigration. "Insurmountable material difficulties of every type," he explained, prevented him from continuing his research in Russia. Furthermore, he was "profoundly convinced that the social experiment being conducted on Russia is doomed to certain failure and that it will yield no result save the political and cultural death of my homeland; this thought oppresses me relentlessly and prevents me from concentrating on my scientific work." Finally, he could not function as "a serf, a slave of others," and so refused to surrender to the state control over his work and its rewards.

Pavlov did not require permission to communicate with his foreign colleagues (he was already doing so through various channels) or even to leave the country. Perhaps he was simply behaving with characteristic dignity and honesty; or perhaps he was opening negotiations on behalf of himself and Russia's scientific community.

Lenin treated his letter as a basis for negotiation. He informed Grigorii Zinoviev, head of the Communist Party in Petrograd, that allowing Pavlov to leave would "hardly be rational," but that he "represents such a great cultural value" that he should not be "forcibly restrained . . . under conditions of material need." During the civil war, the Bolsheviks had encouraged Pavlov to emigrate, but now, on the verge of victory, Lenin viewed him as an important member of a scientific community critical to the tasks of socialist construction.

A state decree of January 1921 ensured Pavlov a privileged place in Soviet science and society. It charged a special commission with creating "most propitious conditions" for his living quarters and central lab, awarded him property rights to all sales of his works, and assigned the Pavlovs a food ration twice the normal academician's allotment. Pavlov initially refused the special ration, insisting that he could not "enjoy privileges not enjoyed by his colleagues." The state understood his position, correctly, not as a rejection of privilege per se, but as a demand for broader assistance to scientists, who were soon provided with state rations (though not so bountiful as Pavlov's).

Meanwhile, his inquiries in the West produced disappointing results. His many admiring foreign colleagues collected money to ease his life and sent meat to feed his dogs; they were even willing, perhaps, to support him modestly in such inexpensive locations as Copenhagen, Stockholm, and Helsinki. They did not, however, offer him an academic position, let alone one with the large-scale support required for his style of scientific research. He was, after all, seventy-two years old, his Nobel Prize–winning work some twenty years behind him, and his conditional reflex research little known in the West.

Pavlov had contemplated emigration only under the most extreme conditions and would certainly not leave Russia for a modest retirement in a foreign culture. His prospects in revolutionary Russia were now much brighter: stability was returning with the end of the civil war, and Lenin himself had made clear the state's interest in his work and well-being.

By mid-1921, then, Pavlov and the Bolsheviks had reached an accommodation. The state would not permit him to emigrate, but would extend special privileges to him and his labs. This understanding, moreover, had been reached while Pavlov criticized the Bolsheviks roundly and insisted that doing so was a matter of personal honor.

Prosperous dissident, 1921–29

In the 1920s, Pavlov became a prosperous dissident—living a materially comfortable life, presiding over a flourishing, state-funded scientific enterprise, and, protected by his special status, the Communists' most vocal, uncompromising public critic.

Yet the same special status that enabled him to openly criticize Soviet power inevitably changed Pavlov's relations to the Communist Party. To manage his scientific institutions and exploit his privileges for the benefit of himself and others, he had to work constantly with the state apparatus. So, during the 1920s, while consistently hostile to the Bolsheviks in general, Pavlov developed a smooth working relationship with various central ministries, the Leningrad Party, and two important Communists, Lev Fedorov and Nikolai Bukharin.

Fedorov had taught experimental psychology and served as political commissar at Tomsk University before the Sovnarkom assigned him in mid-1923 to Pavlov's lab at the IEM. His tasks were to monitor Pavlov and facilitate his research while organizing a Communist presence at the country's leading medical-investigative institution. Over the next few years he became an important party apparatchik while also impressing Pavlov with his dedication and talent for research and management. With Pavlov's support, Fedorov became assistant director of the IEM and editor of the country's leading physiological journal. In 1931, he became the IEM's first Communist director and rose to national prominence, advising Stalin on important scientific–medical matters. Pavlov knew that Fedorov was handling him for the party, but his attitude toward him was respectful and even friendly.

State largesse restored Pavlov's labs to life by mid-1921. The renovated facilities were staffed by a new, much-expanded workforce—many of whose members were attracted by the great prestige of science in Soviet society and by Pavlov's growing reputation. He abandoned one of his three labs when he resigned from the Military–Medical Academy in 1924 to protest its purge of students from clerical families, but that loss was more than compensated by the expansion of his labs at the Academy of Sciences, which now became a Physiological Institute, and the IEM, with its new Tower of Silence. Encased in fortress-like walls and vibration-proof floors, the tower's high-tech chambers enabled experimenters to conduct trials from outside the room and isolated the dog in an ostensibly "neutral" environment.

All this produced Pavlov's first recorded, albeit backhanded, compliment for the Bolsheviks. Chatting with his coworkers upon returning from a short trip to Paris in 1926, he commented on the poverty of French labs and, after a brief reflective pause, added, "Yes, you must give our barbarians one thing: they understand the value of science."

His relationship with Bukharin evolved over the years. Praised by Lenin as "the darling of the Party," Bukharin was in 1924 a leading theorist and member of the Central Committee when Pavlov devoted the first lecture of his course on physiology to a denunciation of Bukharin's *Proletarian Revolution and Culture*. He began:

> I was, am, and will remain a Russian person, a son of the
> homeland; I am most of all interested in its life, I live by its
> interests, its moral dignity fortifies my own. I was not a little
> surprised when … history placed before me the question: would my
> homeland be or not be? … The very thought that my homeland will
> perish would deprive me completely of the basic sense of my
> scientific activity. For whom, then, would I strive? … I live with two
> thoughts. On the one hand, with thoughts about physiology, and on
> the other—what will come of my homeland, what awaits my
> homeland, to what end is all this leading?

Lacerating Bolshevik policies, especially toward science, he accused them of betraying Russia's welfare in pursuit of fantasies about world revolution and excoriated Bukharin's pledge to abolish the "anarchy of cultural-intellectual production" through the same planning principles employed to produce textiles and sausage.

Pavlov's public criticisms made him a national symbol of political resistance. One widespread rumor—encouraged by his protests against the persecution of religion—held (falsely) that he was himself a religious believer. He also loudly denounced the campaign of intimidation in 1928–29 that cowed members of the Academy of Sciences into electing unqualified Communist sympathizers and (as Pavlov put it) "terrorists."

Among those new "terrorist-academicians" was Bukharin, who by this time had lost much of his power in his losing struggle with Stalin. Bukharin believed that Pavlov's worldview provided ample common ground for a conciliatory approach, replying to one member of Stalin's faction that "I know he does not sing The Internationale … but he is the world's *leading physiologist, a materialist*, and, despite all his grumbling, *ideologically* he is working for us." A leading advocate of détente with the bourgeois intelligentsia, Bukharin was instructed to "befriend" Pavlov. He began to do so effectively in 1928–29, "breaking the ice" by demonstrating his erudition and, especially, his knowledge of butterflies (Pavlov had been an avid collector) and gradually convincing the scientist that he, too, sought a more democratic and humane state. As their relationship matured, Pavlov came to identify Bukharin with his hope for the moderation of Bolshevism and spoke of him with great regard. Here was "an intelligent Communist," learned and humane, but an unrealistic dreamer. "I'd like to have him in our dog stand for a few years," Pavlov once remarked; "We would teach him to correctly reflect actual reality."

In December 1929, Pavlov turned a celebration of the 100th anniversary of Ivan Sechenov's birth into a startling political spectacle. Asked to introduce the program with a few words, he strode dramatically toward a large portrait of the "Father of Russian Physiology" and addressed it:

> Oh noble and stern apparition! How you would have suffered if in living human form you still remained among us! We live under the rule of the cruel principle that the state and authority are everything; that the person, the citizen, is nothing. Life, freedom, dignity, convictions, beliefs, habits, the possibility of studying, means for life, food, housing, clothing—everything is in the hands of the government. For the average citizen, there is only unquestioning obedience. Naturally, gentlemen, the entire citizenry is transformed into a quivering, slavish mass…On such a basis, gentlemen, not only can no civilized state be built, but no state at all can long survive.

He was not granted a public forum again for more than five years.

Contradictory sentiments, 1930–36

During the 1930s, Pavlov's opinion of and relationship with the Bolsheviks changed gradually—and, by 1934–36, qualitatively. Having earlier regarded the Soviet state as an unrealistic and impermanent terrorist regime, he came to view it as his country's government, with a mixed record of ongoing crimes, blunders, and important achievements.

The seemingly contradictory direction of Soviet policy in 1933–36 was—even for those who understood Stalinism much better than Pavlov—difficult to evaluate. On the one hand, especially after the Nazi seizure of power in 1933, there were signs of moderation: the more reasonable Second Five-Year Plan and the end of rationing, the loosening of ideological controls in many fields, Bukharin's return to political eminence (though not power), and the promise of a new constitution that would guarantee a secret ballot and other basic democratic rights. On the other hand, in the three months after Leningrad Party chief Sergei Kirov's assassination in December 1934, thousands of Leningraders were swept up by the "quiet terror."

The same fundamental elements of Pavlov's prerevolutionary worldview that had earlier set him against the Bolsheviks—his scientism and state patriotism—militated in the 1930s toward a rapprochement with them. Pavlov had believed since the 1860s that the growth and cultural impact of scientific knowledge was a sure guarantor of social progress. In the USSR, the number of institutions and cadres was increasing at a remarkable rate, and science enjoyed unprecedented cultural prestige. He was also impressed by his Communist coworkers, whom he often scolded about party policy, but valued for their dedication and, in some cases, as "thinking people" in their research. The inevitable consequences of scientific literacy, for Pavlov, were profound: if Soviet youth took their science seriously, he informed students in one lecture, they would abandon Communist doctrine, because "science and dogmatism are completely incompatible." In that same spirit, he confidently informed one Communist official that "whatever new people the authorities promote, if they are not stupid and become educated in the full meaning of that term, they cannot but, sooner or later, recognize the right of freedom of thought." Scientific culture would eventually bring with it realism and balance, and so reasonable, humane, democratic policies.

This scientistic faith encouraged him to look for promising signs during a confusing time in which such signs (like patterns in experimental results) could certainly be discerned. So, for example, he opened a gathering of his coworkers in February 1935 with these hopeful words: "I have complained many times about the oppressiveness of life. Now I want to say something different. It seems to me that our life is changing for the better." The hopeful signs—the "swallows of Spring"—were the end of rationing and the promise of a secret ballot. "I want to believe that there is really occurring a turn toward a normal structure of life."

Pavlov was also preoccupied by the threats of Japanese militarism and, especially, Nazism. So, as when tsarist Russia entered the World War, he was powerfully inclined to support his country's government. Shortly after the Nazi seizure of power, a provincial scientist who had enjoyed swapping anti-Soviet jokes with Pavlov in the 1920s discovered that things had changed when he shared a new one with the chief: "It is a scoundrel," Pavlov snapped, "who undermines his government when the homeland is in danger."

Pavlov's working relationship with the state apparatus also gradually accustomed him to insider attitudes: railing publicly against Soviet power accomplished little, but he could sometimes do some good by using his connections. At the same time, he was also subject to constant surveillance in the lab, on the street, and at home with his family. The authorities used the information gathered to manipulate him—often acting through Pavlov's intimates, especially Petrova and his son Vladimir, who, motivated

5. Pavlov's kingdom, the Institute of Experimental Genetics of Higher Nervous Activity in the village of Koltushi, under construction around 1934. In the foreground is part of the chief's beloved gardens, and in the background is the local church of Peter and Paul, which Pavlov took under his protective wing.

by some combination of fear and conviction, could be counted on to intervene at key moments on the party's behalf.

For Pavlov, the 1930s were also years of unparalleled privilege and prosperity. When he complained that noisy traffic outside the IEM disturbed his dogs, the offending street was moved; when he fell ill and a physician prescribed foreign champagne, it arrived the next day from Helsinki, and when he wanted more space for his Physiological Institute at the Academy of Sciences, the president of the academy was ousted from his neighboring quarters. The USSR's most famous scientist was served by a chauffeured Lincoln (though he usually preferred to walk) and his pantry regularly supplied with imported food.

Most moving for Pavlov was the construction of a massive new research center, the Institute of Experimental Genetics of Higher Nervous Activity, in the town of Koltushi, outside Leningrad. A decree of 1929 celebrated his eightieth birthday with massive funding for this new facility; another in 1934 marked his eighty-fifth with even more lavish support. State largesse created a science village, supplied by its own collective farm, that became a symbol of Soviet science and a second home for an aging scientist with a taste for life in the countryside. This project also brought Pavlov into regular contact with Vyacheslav Molotov, chair of the Sovnarkom and Stalin's closest associate, and with Grigorii Kaminskii, commissar of public health. His profound sense of moral obligation to repay the Russian people and Soviet state for this fulsome support also created an emotional bond with the government.

Although he ceased criticizing the government publicly—and with foreign colleagues, even privately—Pavlov continued to do so trenchantly in private conversation, comments to coworkers, and outraged letters to Communist leaders. In one particularly eloquent letter to Molotov in December 1934, he wrote that "it is difficult, sometimes very difficult, to live here." The attempt to build a new type of society was "grandiose in its courage," yet it was only an experiment; and like every experiment its final result remained unknown. Moreover, "this experiment is *terribly expensive*...with the extermination of all cultural peace and all the cultural beauty of life":

> We have lived and are living under an unrelenting regime of terror and violence....For Man, having come from the beast, to descend is easy but to ascend is difficult. Those who maliciously sentence to death masses of their own kind and do so with satisfaction, like those who are forced to participate in it, can hardly remain beings who feel and think *in a human way*. And the reverse. Those who are transformed into beaten animals can hardly become beings with a sense of their own *human moral dignity*....

Have mercy on our homeland and on us.

Molotov passed the letter to Stalin with the comment, "A new nonsensical letter from academician Pavlov."

Defying official campaigns against the church, Pavlov closed his labs and discontinued his Wednesday conferences during Easter and Christmas. At Koltushi, the Pavlovs marked Christmas with a traditional tree and attended Easter services and supported the beleaguered church morally and financially.

He rejected every attempt by the state to intervene in his own labs. In 1933, when a militant from the Section of Scientific Workers arrived at his Physiological Institute and announced his assignment to purge it of undesirables, Pavlov shouted, "Get out, bastard!," grabbed him by the collar, kneed him in the back, and chased him down the stairs and out the door. When the section informed Leningrad Party chief Sergei Kirov of this outrage, he responded, bearing in mind Pavlov's special status, that "I can't help you."

He used that special status to save many victims from the terror. Some were from his personal and scientific circle, including Stanislav Vyrzhikovskii, the first director of Koltushi; Nikolai Krasnogorskii, a longtime favorite who managed Pavlov's Nervous Clinic at the IEM; and two valued Communist coworkers, Petr Denisov and Fedor Maiorov; as well as members of coworkers' families. He won the release of A. I. Barkhatova, the woman who cleaned the kennels at the Physiological Institute, by insisting that she was critical to his research. Countless desperate others camped outside his apartment or appealed to his son and official secretary, Vsevolod. "There were many such matters," Vsevolod's wife later recalled, and her husband "frequently returned home in a state of moral exhaustion." Pavlov suffered guilt and heart palpitations.

He was, then, hardly unknowing or uncaring about Stalinist crimes; yet his general perspective and attitude had changed, as would soon be evident on the world stage. For years, Communist officials had sought Pavlov's support for holding the International Congress of Physiology in Russia. He had refused in 1926 and 1929, fearing Russia would be embarrassed by rejection of its invitation or, much worse, by the poor conditions that visiting delegates would encounter. In 1932 he consented, now confident that the invitation would be accepted, the state would organize the event successfully, and Russia's new scientific cadres and facilities (including his own at Koltushi) would make a good impression. The congress that transpired in Leningrad and Moscow in August 1935 would provide the dramatic centerpiece of Pavlov's final year.

Chapter 6
Nervous types

Pavlov's dogs were recognizable individuals who lived for years in the lab, displaying full-blown personalities that became an integral part of lab culture, a constant factor in the interpretation of experimental results, and the theme of a central, integrative line of investigation.

The chief and his collaborators routinely describe their canine coworkers as weak or strong, lethargic or active, obtuse or intelligent, compliant or independent, passive or impressionable, aloof or sociable, modest or greedy, cowardly or heroic. Coworker Rita Rait-Kovaleva described her Bes as plodding and unimaginative but a "good worker." Undistracted by extraneous noises and movements, this proletarian animal formed conditional reflexes (CRs) slowly, but his hard-won knowledge proved "solid and lasting." Her other dog, the "scholarly" Toy, was easily distracted and, if not challenged with demanding work, even fell asleep, bored, on the stand. Yet, once engaged, Toy easily outpaced the workmanlike Bes.

When studying digestion, Pavlov had pronounced the dog "almost a participant in the experiments conducted upon it"; in CR research, there was no "almost" about it. Lab animals were not "experimented upon"; rather, they "worked." Some worked well, others poorly. Some tasks put before the dog during its workday—

such as forming a delayed CR to electrical shock—were so extremely demanding that only a noteworthy few proved capable of them. Dogs responded differently to the sacrifice of their freedom for the restraint of the experimental stand, and some bridled at their working conditions—refusing, for example, to cooperate with a new coworker or labor in the isolated chambers of the Tower of Silence. They could be hurt or even broken by the burdens of work or life in general—by an overly taxing task, a fight in the kennel, or the loss of a sexual partner. Sometimes an experience on the stand or with another dog elicited an unexpected reaction, exposing a deep wound from its prelab life. Burdened beyond endurance, it might whine, refuse to eat or work, behave in an uncharacteristically timid or aggressive manner, or present symptoms of mental illness.

The quality of a dog's work and personality was very important for its human collaborator. Particularly intriguing results might attract Pavlov to the bench, leading to a close collaboration, enthusiastic reports about the coworker's labors during the lab's regular and spontaneous gatherings, mention in Pavlov's talks and publications, and perhaps a lab or professorship of one's own. It was only fitting, then, that coworkers were frequently photographed with their dogs, and in 1926 marked the twenty-fifth anniversary of Pavlov's CR research with a jocular greeting from his canine "scientific collaborators." The following year they presented him with a photo album of forty currently employed dogs with a list (but no photos) of each animal's current human partner.

During the first decade of CR research, coworkers frequently observed that different dogs responded differently to identical experiments; by 1918 several tentative typologies had emerged, and in the 1920s and 1930s investigation of nervous types was central to lab research. The task was to explain the differing results of CR experiments and to map these on the dogs' varying behaviors and personalities. Thus, the physiological would be

6. Pavlov's coworkers gave him this jocular greeting from his "canine collaborators" to mark the twenty-fifth anniversary of the first experiments on conditional reflexes. The dog on the far left, "Foolish coward," announces, "I refuse food in protest against the insulting psychological nicknames!" The dog second from the right, "Napoleon," observes, "The master is sleeping—I'll have a little snooze myself!" (Pavlov did sometimes discover coworkers sleeping in the isolated chambers.)

fused with the psychological, and both would be rendered *pravil'nyi*.

This was, to say the least, a very ambitious endeavor. As Pavlov devoted to it more coworkers, dogs, and experimental trials, the data became increasingly discordant and their relationship to personality and behavior increasingly difficult to divine. Conceptualizing his canine collaborators as an animal (or person) in nature and society, he constantly introduced plausible new variables—the onset of mental illness and the interplay of nature/nurture, for example—that might deliver that "wonderful moment when suddenly everything becomes clear." But more data just compounded the problem. The doctrine of nervous types became more far-reaching, but also less precise, less refutable, less *pravil'nyi*.

Here we explore this history through three episodes of the 1920s.

Postrel, Milord, and Freud's Anna O.

In mid-1922, intent on explaining the divergent results in identical experiments on different dogs, Pavlov joined Maria Petrova in their semiprivate experimental quarters on the second floor of the Tower of Silence to launch a three-year comparative study of two dogs, Postrel and Milord, whose personalities and performance in experiments differed markedly.

The older Milord was "solid, balanced, and peaceful"; the younger Postrel was "very lively and active." Petrova and Pavlov first established that the dogs' personalities matched their salivary patterns in experiments establishing CRs to six stimuli. Milord efficiently formed the six CRs and also the more difficult delayed reflex to each. A "successful" delayed CR manifested the standard salivary response to a stimulus, but only after a time lag—and this, according to lab doctrine, required a balance of excitation and inhibition. Milord enjoyed exercising that balance, afterward becoming "livelier in appearance" and eating with gusto. The spritely Postrel formed the CRs even more quickly, but proved inferior at differentiation and delayed reflexes. The "confrontation of excitatory and inhibitory processes" involved in those tasks proved so trying that he "fell into such an irritated state and such a fury that further work was impossible."

Having established that their dogs differed in nervous type, Pavlov and Petrova decided to test what happened when this collision of excitation and inhibition was taken "to the extreme, to the point of a break." This turn reflected Pavlov's long-standing interest in psychiatry. In the 1890s, probably inspired by his family history, his own "morbid spontaneous paroxysms," and his bouts (and those of his friends) with neurasthenia and hysteria in the 1870s and 1880s, Pavlov had begun regularly to make the rounds with clinicians at the Alexander III Home for the Care of the Mentally Ill, where his friend Alexander Timofeev was director. During Russia's civil war, when experimental research ground to a halt, he often skied to the home, and in an article on "Psychiatry as a Partner of the Physiology of the Cerebral Hemispheres" (1919) he compared some of its patients with the lab dog Norka.

He later attributed the idea of "breaking" Postrel and Milord to the influence of Sigmund Freud's analysis of the landmark case of Anna O. Generally critical of Freudian theory, Pavlov credited the founder of psychoanalysis with important insights that, like those of physicians in general, required the undergirding of a truly scientific, physiological analysis. As he put it to American coworker Horsley Gantt, "We cannot ascribe all of his ideas to fantasy. There is much truth in what he says…and this is because it is based on physiology. He makes the mistake of considering himself a psychiatrist, instead of a physiologist, and he uses psychological terminology. There is much reality in what he has to say about the collisions and inhibitions in the brain."

Reminiscing years later, Pavlov explicitly identified Freud as his inspiration to "produce neuroses in dogs by means of collisions." The stenographic account renders his remarks in the third person:

> In one of his early works Freud described a case of neurosis in a woman who had for many years needed to care for her sick, fatally ill father whom she loved very much, and who had suffered terribly from the expectation of his inevitable death, attempting all the while to appear happy to him, hiding from him the seriousness of his illness. Through psychoanalysis Freud established that this lay at the basis of the neurosis that developed later. Viewing this as the difficult confrontation of the processes of inhibition and excitation, Ivan Petrovich immediately proposed using this same difficult

confrontation of two opposing processes as the fundamental method for eliciting experimental neuroses in dogs.

Freud had used the case of Anna O. in *Studies on Hysteria* (1895) to illustrate his view that "hysterics suffer mainly from reminiscences" and to promote the talking cure. His analysis featured psychic forces that conflicted, combined, and inhibited one another, and he sought to explicate the dynamics of these processes while remaining agnostic about their ultimate nature.

Pavlov identified Freud's conflicting psychical actors as his own fundamental nervous forces, excitation and inhibition. Anna O.'s hysteria resulted from the conflict between the powerful excitatory impulse of her grief and the strong inhibitory impulse of her determination to hide that grief from her father. The physiological basis of these emotions was identical in humans and dogs, Pavlov reasoned, so he should be able to reproduce this same conflict in Postrel and Milord.

The experimenters did so by using the same procedure as had Erofeeva a decade earlier: by feeding the hungry dog only in association with an electrical shock—that is, establishing shock as a conditional stimulus (CS). This, they reasoned, created a clash between excitatory and inhibitory impulses in the dogs' nervous systems. When the animal began to salivate in response to shock, Petrova increased its intensity to increase the severity of that clash and so subject Milord and Postrel to the same basic plight as Freud's Anna O.

Each dog eventually "broke"—but in a different direction. Postrel's reflexes "lost their regular character" and experiments showed that his inhibitory process, previously weak, was shattered, leaving him pathologically overexcitable. The better-balanced Milord proved more resilient, continuing to salivate at ever-higher levels of shock, but he, too, eventually broke—in "the direction of inhibition. The dogs now suffered "two different neuroses corresponding to the differences in their nervous system."

Having rendered their dogs neurotic, the experimenters attempted to cure them. Drawing on her clinical experience, Petrova rested and then treated them with bromide salts "in light of its indubitable action in the treatment of several nervous illnesses in people." (Pavlov had himself been treated with bromides during his bouts with neurasthenia and hysteria.) Milord did not respond to either therapy and was retired from experimental practice. Postrel, on the other hand, responded well. The dog's ability to differentiate was restored, and his conditional inhibitors now elicited a satisfyingly perfect zero drops of saliva. That cure, however, proved short-lived.

It now became standard lab procedure to analyze dogs' nervous type, and psychiatric concepts came to play an increasingly important role in the interpretation of experimental results. One result was the reinterpretation of Erofeeva's much-ballyhooed experimental trials of 1911–12: Earlier, her dogs' salivation to a combination of food and electrical shock had led Pavlov to equate them with the Christian martyrs who, inspired by love of God, had borne their torture with a smile. Now they were rediagnosed as broken, mentally ill.

Avgust, the flood, and the physiology of cowardice

On September 24, 1924, a catastrophic flood inundated Leningrad, washing over the grounds of Pavlov's lab at the IEM, where some 100 dogs were soon pressed against the wire ceilings of their kennel, straining to keep their noses above the rising waters. Coworkers rescued them by swimming into the cages, grabbing the terrified animals by the head, forcing them underwater, guiding them to safety through the submerged cage door, and ferrying them to the safety of a distant building. The rescuers noticed that "the animals at this time completely changed their usual behavior. They huddled close to people and to each other; many wailed loudly, and even the most aggressive among them did not provoke fights, but rather were obedient." Even dogs that had been "praised for their stronger nervous organization" were much distressed. But all were saved.

Among the survivors was Avgust, described by coworker Alexander Speranskii as "a very lively, mobile, greedy dog" distinguished by "submissiveness and cowardice." For example, if Speranskii shouted, clapped his hands, or moved suddenly, Avgust would wag his tail, squat on the floor, and urinate. Avgust had nevertheless proved a good experimental animal, developing healthy CRs to six stimuli and performing differentiations successfully. When experiments resumed one week after the flood, however, his CRs had disappeared and he responded to the feedbag with a "clearly negative movement reaction" (that is, fearfully). Over the next few days, Speranskii monitored the dog's health and instructed attendants not to feed him off the stand. Yet the strange behavior persisted.

Pavlov now joined Speranskii and noticed that the dog's affect changed when alone. Peacefully occupying the stand when his master was present, Avgust panted and moved fitfully upon his departure. Exposure to a CS exacerbated this response. When Speranskii remained in the room while Pavlov conducted experiments from outside, Avgust approached and ate from the feedbag normally and responded to most CSs as he had before the flood.

His reaction to the buzzer, however, showed that he had not recovered. The dog "suddenly hopped to his feet and began to fitfully move upon the stand and howl." When the buzzer ceased, he exhibited a "passive-defensive reaction" and responded weakly to all CSs. The experimenters then used Speranskii's presence in the room to reverse this, weaning the dog off this "social factor" by gradually replacing the coworker with "components" of his

7. Coworker Alexander Speranskii poses with the lab's celebrated coward Avgust. The dog's erratic behavior after his near drowning during the Leningrad flood of 1924 launched Speranskii's career and played a central role in studies of experimental neurosis.

presence (for example, his jacket). Avgust now responded even to the buzzer with levels of salivation that approximated preflood levels, though the data revealed some continuing pathology.

Hypothesizing that Avgust's peculiar reactions resulted from the trauma of the flood, Pavlov and Speranskii prepared a decisive experiment. A buzzer was sounded to recreate the alarms of that day and a hose noisily flooded Avgust's experimental chamber. The dog "tossed about against the straps, whined, and strained against the stand! And after this he did not eat, gave no reflexes whatsoever—in a word, 'he broke.'" Speranskii then began applying Avgust's CSs. The animal responded only by panting and twisting in the stand. Every attempt to restore his CRs failed, eliciting "various hypnotic phases intermediate between wakefulness and sleep."

The experimenters diagnosed Avgust with "traumatic neurosis, or fear neurosis," elicited by "a sharp disturbance of the balance of excitation and inhibition." Only a few of the dogs subjected to the flood manifested this neurosis, which testified to the importance of inborn constitution. It made good sense that Avgust—"a dog with a sharply expressed passive-defensive reflex, and characterized by exaggerated inhibitory processes"—had responded by "breaking in the direction" of inhibition. The combination of constitution and experience had rendered him neurotic.

In a speech to the Parisian Society of Psychologists in December 1925, Pavlov used Avgust's ordeal to demonstrate the power of CR methodology to illuminate psychological subjects and to propose a classification of nervous types that joined his research to Western medicine's most ancient tradition. The dogs in his lab displayed a great range of behaviors and personalities, which had long made it difficult "to completely reproduce our facts in different animals." When grouped by nervous type, however, these varied results fell

into place, corresponding to differences in behavior and
personality.

Using Avgust as his model, Pavlov explained the behavior,
personality, and performance in CR experiments of cowardly dogs
as results of the predominance of inhibition over excitation. The
relative weakness of the latter process was probably rooted in the
dog's cortical cells, which contained either an inadequate supply
of excitatory material or low-quality material that was easily
consumed. An excitatory stimulus, then, quickly exhausted such
dogs' excitatory capacity, eliciting an inhibitory impulse that
protected the cells from damage. Avgust's salivary responses
during experiments closely mirrored his behavior and
psychological character. When his CRs diminished, the dog also
responded to food, not by approaching it to eat but rather with a
"passive-defensive," fearful response. Clearly, "at the foundation
of...normal timidity and cowardice, and especially pathological
phobias, lies the simple predominance of the physiological process
of inhibition as an expression of the weakness of the cortical cells."

Surveying the dogs in his labs, Pavlov claimed that they
corresponded to the classical Hippocratic types. For the
Hippocratic physician, each constitutional type (sanguinic,
melancholic, phlegmatic, and choleric) resulted from a particular
combination of humors (blood, black bile, phlegm, yellow bile)
and had its own recognizable predispositions and personality.
When the combination of constitution, environment, and way of
life gave rise to humoral imbalances, this resulted in poor health,
which the physician rectified through bleeding, purging, and
regimen. For Pavlov, the connection between his experimental
physiology and ancient medical wisdom—like that between his
digestive research and physicians' emphasis on appetite—
demonstrated the truth of his research and its ability to explain
the scientific basis of time-honored empirical wisdom.

Pavlov claimed that his findings fit the Hippocratic typology "especially" well for the two "extreme types," the sanguinic and the melancholic. Postrel served as model sanguinic and Avgust as melancholic: "Is it not natural to consider and name him melancholic if at every step, at every moment, the surroundings elicit in him always the very same relentless passive-defensive reflex?" Between these two types were two variants of the "balanced type," presumably, the phlegmatic and choleric. In each, the excitatory and inhibitory processes were of adequate strength and interacted in a "precise and timely" fashion. He offered neither examples nor elaboration.

In his 1927 monograph, Pavlov observed that these intermediary types were "better adapted to the natural conditions of life and therefore biologically more resistant" and conceded that "a large number of animals cannot be placed definitely in any of the four types." Returning with Petrova to the Tower of Silence, he experimented on several such animals (chosen, again, for their behavior and affect), which led to the introduction of a third variable: aside from tending toward either excitation or inhibition, dogs (and people) could be relatively "balanced or unbalanced." This brought Pavlov fully into sync with the Hippocratic foursome: the choleric was unbalanced and excitable, the sanguinic excitable but fairly balanced, the phlegmatic tended toward inhibition, but was fairly balanced, and the melancholic was unbalanced and inhibited. This new schema, however, required rediagnosis of particular dogs: the excitable-unbalanced Postrel was now typed as "choleric," replaced as model sanguinic by two dogs, including the "Napoleonic type" Pingel, who were excitable but too well balanced to be broken.

This typology served Pavlov well in brief presentations and became a staple of news coverage when he became internationally famous in the 1920s, but, as he well knew, it hardly explained the great variation in experimental results and personalities—or the relationship between the two—among his experimental animals.

Recognizing that the life experiences of both lab dogs and humans were endlessly complex, he continued to explore new variables toward that fine day when CR data and personality would all fall into place.

Garsik, nature/nurture, and Pavlov's autobiographical insight

Addressing London's Royal Society in 1928, Pavlov expressed the hope that his research would inform people's upbringing and self-education. "I, for one, while looking at these experiments, have understood much both in myself and in others." He was alluding to ongoing studies of a puzzling group of experimental animals that were complicating his typology with yet another variable and, in so doing, resolving a long-standing mystery about himself. These were dogs whose performance in CR experiments clearly did not match their personality traits. That led Pavlov to emphasize the "plasticity of the nervous system," to investigate the influence of life experience and training, and to distinguish between inborn "temperament" and the mature "character" that resulted from the interaction of nature and nurture.

One such investigation was conducted by coworker Nikolai Vinogradov with Umnitsa, another "weak, inhibited type" who, like Avgust, had "worked well" on the experimental stand until the flood left her unable to produce consistent CRs. Vinogradov's experiments highlighted the importance of two dimensions of higher nervous activity that were acquiring increasing recognition: the role of experience (or training) and that of the "social exciter." As a result of Umnitsa's weak nervous system and associations with the flood, she could not initially form a CR to the electrical buzzer, which elicited an immobilizing passive-defensive reflex. So Vinogradov muffled the buzzer, developed in the dog a CR to this gentler, less alarming stimulus, and gradually worked up to and achieved a CR to the buzzer at full strength. That same approach enabled him to extend the length and intensity of

Umnitsa's workday. Vinogradov also noticed that the dog performed poorly in an isolated chamber, but could work longer and respond more precisely in "the constant friendly presence of the experimenter-master." The social exciter, he and Pavlov concluded, raised the general "tonus" of the higher nervous system, which was particularly salutary for weak nervous types.

Vinogradov's techniques were used to heal the shattered inhibitory processes of Postrel (broken earlier by Pavlov and Petrova), reviving the animal's lab career and necessitating another rediagnosis of his nervous type. In the 1930s, Pavlov thrilled to Postrel's achievements in experiments, regaling coworkers with regular reports about this previously underestimated "hero" with "the nerves of a knight."

Two studies of cowardly dogs further complicated the relationship between dogs' performance in CR experiments and their behavioral and personality attributes. Dmitrii Kuimov's dog Felix was "cowardly and submissive," yet performed "marvelously" in the stand, differentiating easily between the metronome at two different speeds, which was achievable only by those with a balanced nervous system. How, then, to explain the animal's cowardice? This, Kuimov concluded—no doubt at the suggestion of the chief, who was observing his experiments—was "probably the result of the dog's earlier training." Felix's clipped tail, they reasoned, testified to his former master's hope that he would become a show dog. That pampered status had spared him the interactions with other dogs and rigorous demands from humans that might have toughened him up to the level his inborn qualities made possible.

Alexander Ivanov-Smolenskii's dog Garsik seemed similarly paradoxical. He shrank fearfully at the approach of a dog or human—and so seemed an "inhibited type with a weak cerebral cortex." Yet he formed both positive and negative CRs quickly, accomplished delayed reflexes easily, and emerged unscathed from experimental challenges that broke weak dogs.

Intrigued, Pavlov worked closely with his coworker on Garsik's case, and they concluded that these contradictory indications resulted from the animal's upbringing. Garsik had been born and raised in the lab's kennel—"that is, in conditions of unfreedom, in a 'jailhouse' regime"—which had encouraged caution, broken his spirit, and produced submissive behaviors that masked his true nervous character. Here, again, the mapping of experimental results on observable behaviors and personalities required another variable: the animal's life experiences. Garsik inspired the chief to create in Koltushi a center for the study of the nature/nurture relationship.

Pavlov was preoccupied with Garsik in 1928 when he confided to his distinguished London audience that experiments on nervous types had illuminated much about "myself." He clearly identified with the dog's divergence between inborn temperament and mature character—as Pavlov must have considered himself an unlikely nervous type to become a successful scientist. As we have seen, he believed that only a firm balance between excitation and inhibition allowed an organism to differentiate precisely between similar stimuli and, generally, to perceive reality correctly. That same balance, he had explained in his impassioned, pained speeches of 1918, was the physiological basis for the great realism of the English and German types, and so for their successes in the culture of daily life, self-government, and science, while the imbalance of the Russian type—the predominance of excitation—underlay his homeland's woeful performance in those same spheres.

Yet Pavlov himself was hardly a strong, balanced type. Famously explosive, he described himself as "an unrestrained choleric" and a "cycloid" (a "cyclically unbalanced strong type"). He was perhaps describing his own feelings during his legendary outbursts when he stated confidently that if an overexcitable dog could speak, it would report "that it cannot restrain itself from doing what it should not."

How, then, could this unbalanced choleric become such a successful scientist? At the turn of the century, Pavlov had enjoyed discussing his scientific style in terms of the German chemist and philosopher Wilhelm Ostwald's classification of thinkers as "romantics" or "classicists." By the 1920s he naturally employed instead the framework of nervous types. Pavlov's longtime associate Iosif Rozental was no doubt drawing on such discussions when, in an article for *Physician's Gazette* on the occasion of Pavlov's eightieth birthday in 1929, he analyzed the chief's nervous type. Pavlov's difficult life before 1890 had developed in him "strong self-control" that balanced his "inborn excitability." The result was "a personality possessing both powerful excitatory and inhibitory processes." This was the physiological basis of Pavlov's scientific style, which combined a powerful excitatory phase (featuring free and even "fantastic" theorizing) with an equally powerful inhibitory phase (with "all-sided critical analysis" of hypotheses). Devoted to Pavlov and intimately familiar with his temper, Rozental would never have published such a revelation, let alone in a celebratory article, without approval from the chief himself.

Pavlov's abilities and achievements, like Garsik's, then, resulted from the interaction of heredity and life experience. His elliptical comment in London about enhanced self-knowledge referred to this resolution of an autobiographical riddle through his anthropomorphic, and often self-referential, experimental research on animals.

Chasing the horizon

Recognizing the importance of life experience immeasurably complicated the definition of nervous types. And this at a time when the state was investing enormous sums to build Pavlov's science village at Koltushi, where the central goal of investigating nature/nurture and breeding "pure nervous types" required a precise and reliable typology.

In November 1930, Pavlov announced to his coworkers that another grand revision was necessary. He had been mistaken to think that "excitable types are specialists at excitation but bad at inhibition, and inhibitory types the reverse." For example, Postrel, who had earlier exemplified the excitable type, had since demonstrated the ability, with sufficient training, to "develop very strong inhibition." So, Pavlov abandoned the four Hippocratic types based on the relative strength of excitation and inhibition and instead proposed the qualities of strength, balance, and lability. Nerve cells, he explained, have two basic qualities: the ability to perform work and lability (to switch responses quickly; the opposite of inertia). Dogs with strong excitatory processes might also have strong inhibitory processes; their relative balance and lability were separate, independent qualities.

He pondered the situation privately and unhappily. Titling a single sheet of paper "Possible types of central nervous system," he began combining systematically the three qualities that influenced a dog's responses to an experiment (and so defined its nervous type), sometimes provisionally identifying these combinations with a particular dog. He began:

1. Strong, balanced, labile.
2. Strong, balanced, inert excitation and inhibition.
3. Strong, balanced, inert excitation, labile inhibition.
4. Strong, balanced, inert inhibition, labile excitation.
5. Strong, unbalanced with a relative dominance of excitation, labile.
6. Strong, unbalanced with a relative dominance of excitation, inert excitation and inhibition.

Working out the various combinations, he listed twelve possible "strong" types before reaching "13. Weak, balanced, labile (passive-defensive)," and delineating twelve possible varieties of weak dogs, identifying one of the lab's notable cowards provisionally within one of these: "22. Weak, relative dominance of inhibition, inert excitation and inhibition (Umnitsa?)." Then came "25. Middling, balanced, labile," but, now headed for at least thirty-six possible types, and with very few of his 150 dogs included in with any of them, he ended his list with the desultory recognition "and so forth." His journey to the horizon continued.

Chapter 7
Year of climaxes

For Ivan Pavlov, the International Physiological Congress of
August 1935 was a lifelong dream come true. The former
seminarian who been inspired by the "people of the 1860s" to
pursue a life in science was now celebrated by an international
gathering of his discipline and hailed as the Prince of World
Physiology. Speaker after speaker praised his achievements, an
army of current and former coworkers—many now masters of
their own lab—presented the latest in conditional reflex (CR)
research, a steady stream of visitors attended his experimental
demonstrations and marveled at Koltushi, journalists from
around the world pressed for an interview, and he was celebrated
by Russia's grateful leaders.

The emotional impact was all the greater because the congress
was much more than a personal achievement—it was the triumph
of his scientist vision and of his homeland. It would have been
unimaginable during Pavlov's youth, and the headiest of fantasies
in 1914, that Russia, so long a scientific backwater, would proudly
host an international congress for which the state spent
unheard-of sums of money, treating delegates like visiting royalty
and dispatching leading figures to address them, while the
country's newspapers breathlessly covered the gathering, cheering
crowds saluted the visiting scientists, and the world's leading
physiologists from the United States, Europe, Great Britain, and

Japan shared their astonished admiration at the country's new scientific facilities and scientific cadres.

The grim international scene provided additional drama and significance. Having underestimated for years the unique threat of Nazism, Stalin had reversed course in December 1933 and embraced Popular Front politics. The USSR joined the League of Nations, belatedly endorsed it as a force to contain fascist Germany, and committed itself to collective security through multilateral treaties with Great Britain and France. "We do not seek an inch of foreign soil," Stalin announced, "nor will we surrender an inch of our own."

The Soviet state was intent on using this first large international scientific gathering in Russia to showcase its support for science and collective security against the threat from Nazi Germany. "Shock work brigades" had completed or renovated scientific institutions and likely sites for visitors in Leningrad and Moscow (the paint was still drying on Koltushi's new dwellings), and a committee of leading state officials and scientists (with Pavlov as official chair) had for many months carefully planned every dimension of the event.

Pavlov looked like "a medieval saint," recalled American physiologist John Fulton, as he called the congress to order at the stunning Tauride Palace and delivered some brief, but powerful introductory remarks. The simultaneous translation was "excellent," and the English version that "came through the ear phones was so well-timed to the old man's spirited gesticulations that one fancied he was actually speaking English. . . . The most remarkable thing was to see a man of 86 years presiding over a huge gathering—great flood lights pointing at him and movies being taken from every direction—all with the fire and enthusiasm of a man of forty."

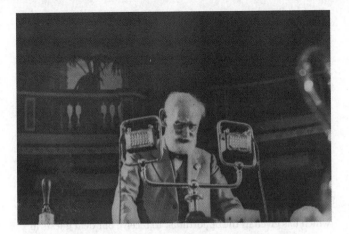

8. Pavlov opens the XVth International Physiological Congress in Leningrad, August 1935. Preoccupied with the threat from Nazi Germany, he announced, "I am happy that the government of my mighty homeland, struggling for peace, is the first in history to proclaim: 'Not an inch of foreign soil!'"

Pavlov's force as personality and symbol combined with the sense of occasion to energize the vast hall. His words were proud, patriotic, and politically pointed. Beginning slowly, he welcomed the delegates and noted the appropriateness of holding the congress for the first time on Russian soil. The gifts of Sechenov memorabilia honoring the "Father of Russian Physiology" should remind visitors that Russian physiology was quite young. Noting the importance of international meetings for popularizing science, he added that in Russia "our state now provides enormously great resources for scientific work and attracts masses of youth to science."

With this first compliment to the Soviet state he pivoted toward his main point: "We are all dear comrades, united in many cases by obvious friendly feelings. We are working, obviously, for the rational, decisive unity of mankind." That was now threatened by war, "a bestial means of resolving vital difficulties, a means unworthy of the human intellect with its immeasurable resources."

These words elicited thunderous applause, after which he continued with a phrase that he had added at the last moment after consultation with Maria Petrova and commissar of public health Grigorii Kaminskii: "And I am happy that the government of my mighty homeland, struggling for peace, is the first in history to proclaim: 'Not an inch of foreign soil!'" Physiologists should be especially sympathetic with this position, and "as seekers of truth, we should add that it is necessary to be strictly fair in international relations. And this is the main and real difficulty." Again applause. He concluded by thanking "our government, which has given us the opportunity to receive our dear guests in an honorable manner." Pavlov had gone well beyond a plea for peace—he had endorsed Soviet foreign policy.

The second night's banquet, in the throne room of the Catherine Palace at the former tsarist residence outside Leningrad, proved equally memorable. "A gargantuan feast had been laid out for 1,400 people," recalled Fulton, who learned that new kitchens had been installed for the occasion and staffed by 80 chefs and 180 waiters from Moscow. The tables were "piled high with elaborate hors d'oeuvres," and at each place were seven wine glasses, a carafe of vodka, and two bottles of wine. When most of the guests were seated, at about half past eight, Pavlov entered to great applause, and the waiters began distributing their delicacies. A hearty ovation greeted the headwaiter when he entered with an effigy of Pavlov in clear ice.

Pavlov's toast, largely inaudible in the din, captured his own deepest feelings about the event. He celebrated natural science as "the main strength of humanity" and physiology as the science that would "teach us how correctly to think, feel, and desire," and so provide "true happiness to human existence." (That last phrase recalls his poem and public speeches during the desperation of

102

spring 1918.) Interviewed by an *Izvestiia* correspondent the next day, he amplified his remarks: Physiology had the special obligation to teach people not only how to work, rest, and nourish themselves correctly, but also "how to think, feel, and desire correctly"—that is, to create a scientific psychology. The Soviet state, he added, understood this: "Take my laboratory at Koltushi. The state has spent and is spending much money on Koltushi; an entire scientific village has been built. And so I now constantly worry how we will redeem these expenditures by scientific work."

Special trains bore the delegates to Moscow for two plenary talks and a final banquet in the Kremlin's Red Hall. Sitting next to Molotov at the head table, Pavlov took the opportunity to inquire skeptically about collectivization and to request that his arrested coworker Adlerberg-Zotova and her family be freed. After Molotov's speech—which sounded the now-familiar themes of peace, collective security, and the cooperative development of science—Pavlov rose with the first toast in response.

He now replayed publicly with Molotov a scene that he had unknowingly "rehearsed" with Kaminskii and Fedorov during their tour of the IEM on the eve of the congress. During that rehearsal, the splendid state of the facilities and Fedorov's rhetorical question ("So, you won't be ashamed to show this to your foreign colleagues?") had elicited Pavlov's now-standard response: the facilities were wonderful, state support for science was bountiful, and he only hoped that he could justify the expenditures. Now, during the final scene at the congress, the overwhelming experience of the past ten days and the task of responding to Molotov's speech served as a powerful stimulus for an identical response: "You have heard and seen the exclusively favorable position that science occupies in our country. I want to illustrate the existing relations between state power and science with just one example: we, the heads of scientific institutions, are anxious and uneasy regarding whether we are able to justify all those resources that our government has made available to us."

Molotov responded as had Kaminskii ten days earlier: "We are certain that you will unconditionally justify this!" Thunderous applause. Pavlov's toast now combined his accumulated sentiments over the past month with the headiness of the moment in his warmest public comments yet about the Soviet state: "As you know, I am an experimenter from head to toe. All my life consists of experiments. Our government is also an experimenter, but of an incomparably greater category. I passionately wish to live in order to see the victorious conclusion of this historical social experiment." To rising applause, he raised his glass "to the great social experimenters."

He had responded to the occasion as he saw fit given the international threats, his optimism about the growth of Soviet science and its inevitable humanizing consequences, and his own gratitude toward the state—all amplified by constant manipulation and his characteristically passionate response to the moment. Yet, as he had pointedly reminded Molotov in his letter of December 1934, the results of the Bolshevik "social experiment" were, as with all experiments, unknown—and this one was "terribly expensive." And the pained entreaties from its victims kept coming.

Pavlov was particularly moved by pleas from those victimized by the campaign against religion. A few weeks after the congress, he received desperate letters from V. A. Gannushchenko and G. A. Bogomolov, who, along with their families, were among the many priests and children of priests who had been exiled and deprived of the right to a higher education and work. Citing Pavlov's remarks at the congress about the special obligation of physiologists to teach "how correctly to work, to think, to feel," Bogomolov implored him to act on behalf of the Leningrad clergy, to bring some peace to "my tormented soul."

Such letters led him to protest this "current injustice that constantly depresses me" in a December 1935 letter to Molotov.

Clerical families had produced such leaders of "vital truth and progress" as Dobroliubov and Chernyshevsky, more than half of prerevolutionary Russia's physicians, and many of its scientists (including, of course, Pavlov himself). "Why are they all included as members of some typologically exploitative class? I am a freethinker and rationalist of the first order and have never been any kind of exploiter, and ... I nevertheless recall my early life with a sense of gratitude both for the lessons of my childhood and for my schooling."

As in earlier responses to Pavlov's letters about repressed individuals, Molotov dismissed his reference to the "many unjustly accused," but assured him that "Soviet power avidly corrects its real mistakes," suggesting he would look into the particular cases Pavlov mentioned. He also informed him that change was coming. Restrictions on the children of the clergy had earlier been "necessary," but now would be lifted. Here Molotov was providing advance word of the Politburo's impending decision to lift the prohibition against children of various outcast groups receiving a higher education. That news gladdened Pavlov—who saw it as another hopeful sign of increasing reason and moderation—but, as he well knew, this hardly ended the repression of religion.

He had told his family for months that he wanted to "do something for religion" and now informed Molotov of his intention to send him a "principled and lengthy" statement about "our state atheism." This, then, was the origin of one manuscript that he would labor on while bedridden with what proved a fatal flu during his final winter at Koltushi.

Eugenics, primates, and reassessment of the conditional reflex

Pavlov's original plan for Koltushi was, as he informed Molotov, to study the dynamics of the nature/nurture relationship, "to determine experimentally on animals the conditions for acquiring by means of selective breeding, a greatly improved nervous system"—"for the use and glory first and foremost of my homeland." That research, he informed a reporter, "should lead to the success of eugenics, the science of the development of a better human type."

For Pavlov, like many geneticists in the 1920s and 1930s, eugenics was the logical practical extension of genetics and, as for thinkers of almost every ideological stripe, a rational application of science to improve medicine and society. Pavlov envisioned a eugenics that would encourage scientifically informed voluntary individual behaviors to compensate for hereditary weaknesses and benefit one's offspring. This made especially good sense for him given his long-standing visceral belief in the inheritance of acquired characteristics (the subject of experiments at Koltushi by his close associate Evgenii Ganike). He condemned Nazi eugenics and observed repeatedly that neither current knowledge of heredity nor "the sensitivity of cultured man" justified mandatory sterilization.

This grand vision, then, integrated several basic interests: the interplay of nature/nurture, the nature and improvement of the "Russian type," and humans' individual and collective human struggle against *sluchainost'*, for control of their own destiny. As he put it in a private note, "Fate, the social struggle with it. Practice. Eugenics."

Research at Koltushi began in fall 1928 when coworkers Stanislav Vyrzhikovskii and Fedor Maiorov divided two litters of four puppies each into two groups of "imprisoned" and "free" dogs. The former remained always in the kennel, whereas the latter roamed freely, interacting with nature, other animals, and people. After about one year, the experimenters compared their behavior, personalities, and responses to CR experiments. For Pavlov, the sharp differences in the performance of the two groups confirmed his analysis of Garsik (and himself)—life experience mattered.

The free dogs were enthusiastic, affectionate, and active; the prisoners were "pathetic cowards" that failed to engage with their environment. (Among people, too, he noted, life in "a hothouse setting" often left those with a strong nervous system in a craven state). Interestingly, prisoners generally performed better in CR experiments, though free dogs more quickly accomplished differentiations.

This research, however, soon bogged down. The building of special kennels proceeded slowly, and further experiments awaited the lengthy process of breeding successive generations to obtain "pure types." Furthermore, Pavlov admittedly lacked a "precise approach" to identifying the hereditary factors for which to select. Most important, he could not breed pure types without a reliable typology. At a meeting in September 1933, he announced that "the doctrine of types must be placed on firm foundations," but soon admitted that "everywhere we are seized by hesitation about the type of nervous system of this or that dog." He collaborated with Maiorov on a set of twenty-one tests for nervous type—but these yielded contradictory diagnoses for many dogs and mapped poorly on their behavior and personality.

Yet Koltushi had moved to the very center of Pavlov's emotional life as a scientific center, a country home, and the repository of dreams for his remaining years—so, it proved quite fortuitous when, in summer 1933, the arrival of two unexpected guests, the chimps Roza and Rafael, provided a new, engrossing scientific rationale for his sojourns to the science village.

That arrival was neither Pavlov's idea nor happy chance: it resulted from the scientific–ideological agenda of the Communists in his lab. They hoped that exposure to primates might convert the chief from the reductionist, mechanistic views he had imbibed in the 1860s to a more dialectical conception of higher nervous activity and might encourage his recent tendency to complement his long-standing "narrowly analytic" focus on the dynamics of individual reflexes with attention to the systemic qualities of the cortex that separated dog from human. As they knew from experience, arguing philosophy with him was useless, but, as a committed and observant experimenter, Pavlov could be moved by "the spontaneous dialectics of facts." Engagement with primates might, then, convince him of the qualitative differences between them and lab dogs.

Pavlov's Communists had for years attempted unsuccessfully to lure him to the Sukhumi Primate Station, where several of them were conducting CR experiments on the synthetic qualities of its inhabitants. Since Muhammad would not go to the mountain, Communist coworker Petr Denisov brought the mountain to Muhammad. Dispatched by the Academy of Sciences to Serge Voronoff's primate colony on France's Côte d'Azur, he returned with the two chimps as a gift for the chief.

Roza and Rafael engaged Pavlov's curiosity. He began immediately to observe and experiment on them in collaboration with Denisov and, beginning in fall 1933, did so during regular Friday sojourns to Koltushi and lengthier stays during summer and winter breaks. This first sustained contact with a model organism other than the dog would indeed have profound consequences for his ideas about CRs and psychology.

Lured into the chimps' arms, Pavlov had his own, long-maturing reasons to embrace them. He was preoccupied with the challenge to associationism by Gestalt theorists, led by Wilhelm Köhler, who made effective use of his experiments with chimps. Gestaltists argued that the human mind engages the world not by accumulation and association of small perceptions, but rather by grasping wholes, configurations, or *Gestalten*. The key concept here was *insight*—the active apprehension of the structure of an entire "perceptual field" that lurked beyond simple experiences. After hearing Köhler's address to the International Psychology Congress in 1929, Pavlov began reading his *Mentality of Apes*

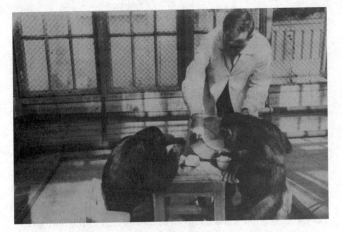

9. Petr Denisov with Roza and Rafael at Koltushi. His collaboration with Pavlov on studies of the chimps' learning process led Pavlov to reconsider his long-held view that the physiological conditional reflex and the psychologists' association were synonymous.

(1917) and *Gestalt Psychology* (1929, 1933), criticizing them with increasing intensity as Gestalt's popularity grew and its advocates targeted his own CR-based associationism. The first year and a half of Denisov and Pavlov's experiments with Roza and Rafael were based largely on those that Köhler had described in *Mentality of Apes*, as Pavlov wanted to examine Köhler's evidence for ascribing to chimps insight that was inexplicable in associationist terms.

In 1930–32, Pavlov had developed his concepts of *systematicity* and the *dynamic stereotype* to provide an associationist explanation of phenomena that the Gestaltists attributed to the mind's active, holistic insight into the structure of an entire perceptual field. Two basic experiments underlay these concepts: If a series of CRs was established in the dog (the buzzer as a conditional stimulus, a flashing light as a conditional inhibitor, and so forth), any variation in the order of those exciters changed its response to each. The animal, then, responded not just to a single stimulus but also to the system of stimuli as a whole. A second experiment demonstrated that once this series was established, the first conditional stimulus alone elicited the same series of salivary responses as did exposure to the entire set. Only after a few such trials did the new circumstances elicit a change in these responses, an alteration in the "dynamic stereotype"—Pavlov's awkward term for expressing that this system was changeable yet durable, exercising an inertial influence of the whole over the parts. Systematicity and the dynamic stereotype, then, were Pavlov's associationist alternatives to the Gestaltists' perceptual field and "active subject."

Pavlov had earlier worked almost exclusively "from the bottom to the top," from the parts to the whole, from the dynamics of isolated individual reflexes toward a characterization of higher nervous activity, behavior, and the psyche. Now he sought also to move from the top to the bottom, from the whole to the parts, from the dynamic stereotype to its component individual reflexes. Earlier, he had compared unconditional reflexes to a direct line between two telephones and CRs to a temporary connection established through a switchboard (the cerebral cortex); now his attention was shifting toward the switchboard itself.

This change was reflected and facilitated by a new metaphor. The long-standing image of "chains of reflexes" remained, but was increasingly subsumed within that of the cortex as a "mosaic" with the glittering light of excited cortical points, the darkness of inhibited ones, and the grays formed by the traces of past experiences. The mosaic metaphor, which first appeared in Pavlov's brief discussion of systematicity in his monograph of 1927, made it easier to envision a different kind of nervous connection: not just that between a subcortical unconditional stimulus and a cortical conditional stimulus, but also multiple linkages between various cortical points of the mosaic—for example, between two indifferent stimuli (such as a light and a metronome that were unconnected to any unconditional reflex and so had no status as signals).

Innumerable stimuli constantly bombarded the cerebral hemispheres, Pavlov now emphasized, and these emanated both from the external world and from the animal's own "internal milieu." This latter addition reflected his interaction with Polish Pavlovian Jerzy Konorski, who impressed upon Pavlov the existence of a second type of CR, which he termed *motor conditioning*, as opposed to Pavlov's *classical conditioning*. (Skinner would later describe Konorskii's motor CR as "instrumental" or "type 2" conditioning.) When Konorski simultaneously fed a dog and moved its leg, a CR was formed and the dog eventually moved its own leg toward that same goal. As Pavlov put it in 1932, "There is this everyday fact, reproduced by us in the laboratory—the formation of a temporary connection between any external excitation and passive movements, with the result that the animal makes active movements in response to certain signals."

These developments encouraged Pavlov, as an unwavering believer in both determinism and free will, to think that the two could be reconciled, that free will could be reconceptualized as a result of the "summative activity of the cortex," which humans could learn to alter through the "direction of energies." He tested that hypothesis in introspective experiments on his own ability to control the otherwise unpredictable direction of his thoughts and desires.

This, then, was the context for Roza and Rafael's arrival at Koltushi. First repeating Köhler's experiments and then devising his own led Pavlov both to deepen his critique of Gestalt's "mysticism" and, in synergy with his earlier ideas and experiments concerning systematicity, to re-examine and then fundamentally amend his own views. While rejecting Köhler's argument that anthropoids displayed a qualitatively different form of intelligence

111

than that of dogs, Pavlov now recognized important differences between the two organisms that influenced their intellect and learning behavior. At one of his regular Wednesday gatherings with coworkers in September 1934, he noted that the anthropoids' four limbs allowed them "to enter into very complex relations with objects" and so to form "a mass of associations, which does not exist in other animals." The chimp was an "ideal tightrope walker," and, watching it "confirm empirically" the stability of stacked boxes, one saw the complex interplay of its multifold associations—"tactile, muscular, visual, and so forth." Whether roaming Koltushi freely or tackling experimental tasks, Roza and Rafael constantly displayed the multidimensionality of their interaction with the environment—simultaneously manipulating, inspecting, and smelling objects (and people) in their domain.

Pavlov soon realized that, regardless of its species characteristics, the experimental dog harnessed to the stand and exposed to one stimulus at a time could only manifest *some* of its qualities. Constrained by experimental design, that dog embodied his long-standing model of the organism responding serially to single stimuli. The unconstrained chimp, however, was free to express its powerful "investigative instinct" simultaneously with eyes, ears, nose, and hands and so embodied the emerging model of systematicity and the multidimensional mosaic.

In early 1935, Pavlov gathered his thoughts in a never-completed manuscript on *The Intellect of Anthropoid Apes*. Here he confidently repeated his comments at the Wednesday meetings about Köhler's fallacies and the explicability of chimps' problem-solving process in terms of associations.

He was much less confident, however, doing so in terms of CRs. This was a crucial distinction. For Pavlov, a CR was the temporary connection between a subcortical and cortical point, or (as he sometimes put it) between an unconditional exciter of the subcortex (such as food, pain, and sex) and a neutral stimulus (such as a buzzer, light, or metronome) that became a signal for this subcortical exciter by simultaneously exciting a point in the cortex. These specific physiological features underlay the laws governing the CR and its biological role as the mechanism through which organisms adapted to a changing environment. *Association* was a looser, more general term—a reference to any connection between two or more experiences, sensations, or images. Pavlov had long argued that the physiologist's CR and the psychologist's association referred to the objective and subjective aspects of the same process. But now he was not so sure.

In *The Intellect of Anthropoid Apes*, he analyzed the "elementary associations, or knowledge, or ideas" involved when the chimp piled boxes to reach a high-hanging fruit. He identified ten, describing some in more detail than others in this rough draft intended for his eyes only. Here, in paraphrase, is his list, with direct citations (in italics) of some notes to himself: (1) Boxes must stand exactly under the fruit; (2) They must be placed one upon the other. *But how?*; (3) The chimp crawls onto the box to test its stability; (4) It moves a second box near, places it on the first box, and tests for stability by crawling on the boxes. *This is a kinesthetic association*; (5) When a number of boxes are piled up, it tests for stability by inspecting the boxes, and *instead of the kinesthetic association, there is another visual association*; (6) Over time, the chimp looks at the structure from a greater distance to see if it is of sufficient height; (7) The chimp tests the height by climbing on the boxes; (8) If rectangular boxes are used in the trial, the chimp might reach the necessary height without using all the boxes; (9) If the boxes are not rectangular, the chimp needs to use all six boxes; (10) If boxes of various sizes are used in the trial, the chimp must undertake the difficult task of stacking them in the correct order, a process that is *probably governed by sight*.

There was no need, then, to resort to Köhler's insight or other (as he often fulminated to coworkers) Gestaltist nonsense. In this "visible and indubitable act of thinking...there is nothing other than simple and complex associations." But what was the nature and mechanism of these associations? As Pavlov wrote in the margin of his manuscript, *"What is united with what?"* Pavlov referred to "a lawful chain of these associations, that is, the association of associations," but what he described was not so much a chain as an interconnected mosaic. "The elementary [associations] are united among themselves in the most varied way," giving rise to "a system of nervous processes that with repetition is implemented more and more easily, becoming more and more fixed."

The key problem, again, was "What is united with what?" The "association of associations" he described differed fundamentally from his old definition of the CR as a temporary connection between a subcortical and cortical point, a signal established through simultaneity. Rather, he repeatedly invoked the connection between, for example, the chimp's kinesthetic and visual associations. He could explain the chimp's complex interaction with its environment and its progress toward a high-hanging fruit by the association of associations, systematicity, and the cortical mosaic. But this explanation drove a wedge between the allegedly synonymous association and CR.

Having pondered this problem for six months while traveling and presiding over the International Congress of Physiology, in November 1935 he dropped a bombshell on his coworkers: contrary to three decades of lab doctrine, CR and association were not synonymous. The CR was but one *type* of association. This raised new, pressing research questions: What are the different types of association and their laws? What are the dynamics and mechanisms of systematicity? Closing the meeting, he left his stunned coworkers with a question that they had long considered closed: "What is a conditional reflex?"

Chapter 8
Final reflections

During his last winter at Koltushi, Pavlov grappled with the lessons of his climactic year in two manuscripts. So different in language and genre—one a scientific paper on "Psychology as a Science," the other a letter to Communist leader Vyacheslav Molotov about Bolshevism, science, and religion—each was a final expression of the same long quest.

Pavlov devoted that scientific paper to "the physiology of associations." Associationism's success in connecting subjective phenomena such as words, thoughts, and feelings with measurable objective actions represented the "most important…achievement of psychology as a science." That achievement was now threatened by Gestalt, which appealed to the animistic belief in a "special substance" separating mind from the rest of nature. Gestaltist interpretations mystified phenomena that were scientifically explicable through his own research on "the objective phenomenon that corresponds to the psychologists' 'association.'" An audience expecting that phenomenon to be the conditional reflex (CR) was in for a surprise.

Pavlov now invoked (and reinterpreted) experiments with dogs that he had initiated with his turn to systematicity. These adopted a design that he had avoided when preoccupied earlier with "chains of reflexes," but which had been developed by other researchers schooled in his methods. In the mid-1920s, two of Pavlov's disciples interested in psychiatry and child development, Krasnogorskii and Ivanov-Smolenskii, had demonstrated that humans routinely formed connections (associations) between two cortical points—that is, between two indifferent stimuli, even if they were not reinforced by a subcortical exciter. For example, when a child was exposed simultaneously to light, a buzzer, and a metronome, and the light was then established as a conditional stimulus for food, the buzzer and metronome also elicited a food response. This was the same type of non-CR association that Pavlov had later observed in experiments with Roza and Rafael.

From October 1932 through early 1934, Pavlov's coworkers I. O. Narbutovich and N. A. Podkopaev conducted similar experiments (using movement rather than salivary reflexes) with the same result. By simultaneously exposing a dog to a rotating figure and a tone, they established an association between the two. Then they established the rotating figure as a conditional stimulus for electrical shock. Now the tone alone, too, elicited this same defensive reaction. Pavlov had earlier interpreted all these connections as CRs, but now, in "Psychology as a Science," he distinguished between the specific CR and the generic association. The CR between rotating figure and shock had been joined to the association between figure and tone in a mosaical "association of associations."

This new metaphorical perspective explained his decades-long failure to establish experimentally a second- and third-order CR that, as the equivalent of the psychologist's associations, combined end to end in long chains to produce complex behaviors, thoughts, emotions, and knowledge. The lab had first attempted in 1905 to form a second-order CR—by establishing, say, light as a signal for food and then a buzzer as a signal for the light. The buzzer did not, as they excitedly expected, elicit salivation. "We were grief-stricken," Pavlov later recalled. He had concluded reluctantly that, rather than forming a second-order conditional stimulus, the buzzer had become a conditional inhibitor. He returned to this problem repeatedly in subsequent decades, but remained unable to form chained reflexes "analogous to our associations." Now, finally, he could explain that troubling mystery: the physiological dimension of the psychologist's linked associations was not a "chain of CRs," but rather a set of CR and non-CR associations that formed independently and became linked in the cortical mosaic.

In "Psychology as a Science," Pavlov explained the physiological differences between these two types of association. The CR involved the linkage of a subcortical point (excited by unconditional drives for food, safety, and sex) and a cortical point that was excited simultaneously by an external exciter. In a non-CR association, "certain kinesthetic, tactile, and visual excitations from certain external objects and their position become connected with other, also visual [excitations] and perhaps also with kinesthetic excitations from one or another external objects." These were linkages between two cortical points. Another difference followed from this: the formation of any association required cortical *tonus* (energy, excitability). In the CR, that tonus was provided by the involvement of the subcortex, by the energy of primal drives. Since non-CR associations did not involve the subcortex, they required an alternate energy source, which Pavlov's lab supplied in experiments by using wavering stimuli to maintain an *orientational reflex* (the physiological dimension of curiosity).

These two types of association produced fundamentally different kinds of knowledge. By forming a CR, the animal learned about the "relationship of separate objects in the environment to *itself*"—for example, that a buzzer signaled an opportunity to eat and a rustling bush the danger of a predator. Through non-CR associations, organisms learned about "the relationship of external things *among themselves*." The chimp stacking boxes learned that a particular visual image of the boxes corresponded to a stable

117

structure. Non-CR associations, then, represented the "embryo of science."

Narrowing the centrality and scope of the CR, he had expanded dramatically that of Pavlovian method and explanation. He was already preparing his lab to study systematicity and all associations—and associations of associations—in the same manner as he had analyzed CRs, and thus to expand his research into new areas of psychology such as the perception of musical sounds and melodies, of form and relationship.

At age eighty-six, Pavlov remained both true to a lifetime's mental habits and remarkably open to changing his mind.

Science, Christianity, and Bolshevism

Ivan Pavlov

Pavlov also labored that winter over his "principled and lengthy" critique of official atheism for Molotov, but was dissatisfied with it and decided instead to write a letter; he never completed that either. These drafts remained in his personal papers.

He began on a positive note: "I must admit that the longer your regime exists, the further it departs from the extremes from which it began, making room for actual reality instead of theoretical constructions. The depersonalization of the human being in the extreme form of Communism and a despotic dictatorship is giving way subtly to a gradual recognition of the rights of the individual." As evident in the reference to the Bolsheviks' improving grasp of "actual reality," his optimistic interpretation of contradictory indications reflected Pavlov's faith in the civilizing influence of science.

Yet they continued their "barbaric" persecution of religion. Freedom of belief, like freedom of scientific inquiry, Pavlov insisted, was a basic human right, a matter of individual conscience. Furthermore, religion played a positive cultural role

10. Ivan Streblov's intimate 1932 portrait captures a contemplative, even vulnerable, side of Pavlov that, though rarely evident in public, was an important part of his persona. Pavlov kept this painting in his study at the Institute of Experimental Medicine.

by relieving the most painful dimension of life—the lack of control over one's fate in the face of the uncontrollable and unexpected (*sluchainost'*).

119

Dismissing the view that religion originated in deception by self-interested exploiters, he attributed it rather to a purposeful, reflexive, adaptive response by "naive man" to the dawn of self-consciousness and man's juxtaposition of himself to nature. Religion was rooted in a fundamental psychological need: oppressed by the *sluchainosti* of his natural and social milieu, man "needed to believe in some law of nature, in some more or less constant connection between cause and effect," so he could assume some control over his own destiny and "depend in a human way upon his own activity." That need was filled by the concept of a god who "held everything in his hands and who, if well disposed toward you, did not permit any evil *sluchainosti*."

Rationalists, Pavlov continued, dismissed religion as a remnant of prerational thought and believed that science, the fruit of human reason, would "completely replace, eliminate religion." That had long been his own view and had played an important part in his decision to leave seminary for science. Now, however, although he remained a rationalist "to the marrow of my bones," he found it unconvincing.

Here his essay took a psychologically revealing turn as he dwelled on the horrors of uncertainty and the uncontrollable in human life. As so often when he addressed broad issues, the autobiographical dimension of his thinking is palpable as he slips constantly between the first and third persons: "What is the most difficult, really terrible thing in life? *Sluchainosti* of birth—the inheritance of genes and...social class; of environment, initial conditions...of death [and]...illness...of every other misfortune and obstacle in life." A peaceful, fulfilling life required "a regular, undisturbed course of life and certainty in it. But where can one obtain either?"

Rationalists replied that science provided an escape from *sluchainosti*—and had succeeded in doing so in many realms—

but (as Pavlov knew from experience) it remained "almost powerless" as an agent of certainty in any individual human life.

> However much I were to conduct myself consciously according to the rules of science, could I really be certain that some unexpected serious illness would not swoop down unexpectedly upon me with various consequences? Although I always walk on the pavements and am careful at all intersections…that a truck will not strike me or a mass of concrete not break off. And my sense of peace is connected with the fate of my intimates…and all such serious *sluchainosti* [in their lives] also shake my internal world. And the fate of my homeland? A mass of *sluchainosti* not even considered by any science [threaten it], to say nothing, of course, about dangers that are rationally foreseeable.

Here Pavlov voiced the preoccupation with certainty and control that had always resided at the heart of his life and his science, fusing his personal psychology and passions with his training in physiology and the scientific spirit of his age. And he was clearly reflecting on his own long life: the tragedies of war and revolution, the recent death of his son (a "bolt from the blue" for his wife; the "*sluchainost'* of illness" for him), his constant distress about the repression of innocents and the "death of my homeland." Perhaps he was also contemplating his mortality and the ultimate *sluchainost'* of death ("You dream about this and that, and suddenly you die!").

His uncharacteristically sober assessment of science's ability to conquer uncertainty seems also a tacit recognition that, for all his achievements over the past three decades, his own research had failed to confine the psyche within the comforting certainties of mechanistic law—and, indeed, seemed nowhere close to doing so. He still believed that science would finally succeed in understanding, limiting, and perhaps controlling the torments of human consciousness—but, as he had written five decades earlier when contemplating Dostoevsky and the prospects for a true science of the psyche, "not soon, not soon."

As it had since time immemorial, religion provided the comforting certainty of a kind, omniscient divinity, a necessary "lightning rod" against the uncertain and unexpected—not just (as he had thought previously) for "weak types," but for the "great majority" of people. Granted that no god existed to protect the believer from the "arbitrary power of unthinking and cruel *sluchainost'*" (which was governed by the "implacable laws of nature"), religious faith still cushioned the faithful against life's unexpected blows that might otherwise destroy one's energy and interest in life.

Religion also contributed to certainty and control of one's destiny by establishing moral ideals of behavior. He was no doubt thinking of Serafima and her response to the recent loss of Vsevolod and perhaps was recalling the comfortable certainties of his years as a young believer: "In order that an omnipotent God protect you from *sluchainosti*, you must please him,…aspire to resemble him,…you must approach the ideal. And when you are afflicted by misfortune, this is either a test of your faith in god or a reminder that you are failing to fulfill his wishes and so must pull yourself together." Religion thus replaced "the unrestrained arbitrariness of external forces" with "an ideal of behavior."

That moral ideal was personified by Jesus, "the apex of humanity—who embodied the greatest of all human truths, the truth of the equality of all people, which provides the basis of the rights of the individual and a moral concept that divides all human history into two halves: the slavery era before Jesus and the era of cultural Christianity after Jesus." What objection could any exact science, or any rational state, have to the moral teachings of this Christianity?

Indeed, here resided a promising commonality between Christianity and Communism. Pavlov acknowledged the "undeniable service" of Communism's moral core: its insistence on "elimination of the distinction between the wealthy and the poor" based on the recognition that useful labor should entitle a person to "respect and welfare in life." By its commitment to "the greatest of all human truths, the truth of the equality of all people," Communism shared considerable ground with Christian culture. This might have become a key subtheme of his essay, had he completed it. As his draft lost structure toward the end, it became a series of scattered comments and thoughts for further development. One of these read, "You are the successors to Jesus's mission." If not for the persecution of religion, "there would be among you fervent and talented adherents from among the servants of the church."

He had twice recently expressed that same sentiment. At a meeting with coworkers after Easter 1935 he commented that one need not celebrate Easter because "Jesus Christ was a man" and could not be resurrected. "But one must necessarily celebrate Christmas. Jesus Christ was a great man. He was the first Communist on earth." Some months later, he remarked that the results of the Bolshevik experiment were unclear and he did not endorse their policies, but "at the basis of Bolshevism lies the striving of the Russian spirit toward perfection, justice, good, toward a great humanity. Karl Marx created this system, but the Russian spirit recast it in its own way.... Bolshevism is more multifaceted and complete than Christianity, but one must still await the results for decades, a half century at least."

Pavlov had become neither a believing Christian nor a convert to Communism, yet he was groping hopefully toward his own grand reconciliation. He had abandoned religion for science and scientism, but now science, perhaps, was producing a more realistic, civilized, and humane Bolshevism that might become a genuine Russian contribution to the great historical era inaugurated by Jesus. Time, he concluded, would tell.

Chapter 9
Epilogue

Pavlov died of pneumonia on February 27, 1936, and was honored at a massive state funeral, where the Communist Party, now unobstructed, began to propagate a Stalinized image of his science and his transformation from critic to convinced supporter of the "Soviet experiment."

The contradictory indications of Pavlov's final years presaged not "the swallows of Spring," but the horrors of high Stalinism. Many of Pavlov's close associates—including Communist officials Bukharin and Kaminskii and coworkers Denisov and Vyrzhikovskii—were arrested and shot. Many others were arrested. Fedorov, Petrova, and Molotov survived and thrived.

Despite substantial criticism of Pavlov's digestive research—especially by those, like William Bayliss and Ernest Starling, who championed humoralist mechanisms as opposed to Pavlov's nervism—he was broadly acknowledged for establishing the basic framework for studies of the digestive system as a coordinated system. For decades thereafter, experimental physiologists disagreed about the relative importance of nerves, hormones, and psyche in the "neuropsychoendocrinological complex." The precise, lawful characteristic secretory curves so important to Pavlov vanished from the scientific literature as organ physiology fell out of fashion.

During the polarizing conflict between behaviorists and Gestaltists, representatives of each falsely identified Pavlov as a behaviorist. His attention to consciousness, personality, and psyche was largely buried under Stalinist dogma and, for anglophones, by the inaccessibility of Russian texts and by incorrect, highly misleading translations.

So, after World War II, when many Western psychologists again turned their attention to animals' inner emotional and cognitive life, Pavlov seemed hopelessly old-fashioned. Yet his scientific practices are compatible with contemporary scientists' attention to personality types among animals, their attempts to divine the goings-on in the minds of dogs and dolphins, and the mechanistic and anthropomorphic metaphors they employ to do so. Pavlov's attempt to integrate physiology, psychology, and psychiatry has fared poorly amid modern hyperspecialization, and his organ physiology has been largely abandoned amid the reductionist drive of modern medical science. Yet that legacy also continues to inspire many scientists in both Russia and the West—for example, at Pavlov's former Institute of Physiology in St. Petersburg and among members of the United States' Pavlovian Society, with its journal *Integrative Physiological and Behavioral Science*.

Pavlov's surgical operations and experiments on animals were not exceptionally cruel for his day, and he was able to shrug off the objections of antivivisectionists (who protested most energetically during his visit to London in 1935). Much greater (though hardly all-inclusive) sensibility about animals' feelings and rights renders many of his procedures unacceptable for many people and institutions in the early twenty-first century.

Physiologists and psychologists today would view Pavlov's map of higher nervous processes as hopelessly naive and oversimplified, and their use of error bars in curves codifies their abandonment of his notion of precise, mechanistic determinism. Few, if any, would agree that the psyche can be explained on the basis of nervous reflexes. Many scientists in both Russia and the West, however, fruitfully employ variants of his conditional reflex methodology in studies of simple learning; fear, anxiety, and stress; the regulation of blood pressure, body temperature, and the immunological system; drug and alcohol addiction; memory; and psychopathology.

Whereas Pavlov scrutinized saliva drops in his attempt to understand the psyche, we can all witness today the oxygenation of neurons on functional magnetic resonance imaging as a monkey reaches for a banana or a person experiences love, hate, or fear. It is much more vivid in our high-tech age, but the problem of translation remains. Modern technology has provided various candidate replacements for Pavlov's metaphors. Computers have displaced factories in cultural consciousness, so the language of feedback loops and hardware/software seems to us more sophisticated than Pavlov's mechanistic mapping of inhibition onto cowardice, trace reflexes onto memories, and changes in the dynamic stereotype onto feelings of unease. Yet the hard question of consciousness—the relationship between body and mind—remains a mystery, perhaps awaiting metaphors beyond our current experience.

Had Pavlov indeed lived to be 100, his scientism would have been powerfully challenged by Nazism and Stalinism, World War II and its murderous finale at Hiroshima and Nagasaki. In our time, too, breathtaking scientific achievements coexist with—and are often deeply implicated in—spiritual impoverishment, mass oppression and destitution, war, and the suicidal destruction of planetary ecology. The social impact of science is what humans and our societies make it.

As in Pavlov's day, science offers illuminating insights into our behaviors and moods and palliatives for our pains, but the challenges of uncertainty and the existential issues of "our consciousness and its torments" that animated his grand quest remain always with us.

References

Three basic sources for this book are I. P. Pavlov, *Polnoe sobranie sochinenii*, 6 vols. (Moscow: Akademiia Nauk, 1951–52) [PSS]; *Pavlovskie sredy: Protokoly i stenogrammy fiziologicheskikh besed*, 3 vols. (Moscow–Leningrad: Akademiia Nauk SSSR, 1949) [PS]; and the archival materials in the Pavlov collection of the Archive of the Russian Academy of Sciences, St. Petersburg branch [ARAN, *fond* 259].

Unless otherwise noted, translations from the Russian are mine.

Chapter 1: Winter at Koltushi

"What…is the most difficult, really terrible thing in life?": ARAN, *fond* 259 *opis'* 1a *delo* 39. [Hereafter ARAN 259.1a.39.]

Chapter 2: Certainty: religious and scientific

Information about Pavlov's youth is from the archive at the Memorial Museum–Home of Academician I. P. Pavlov in Riazan; about the seminary from the State Archive of the Riazan Region; and about Dostoevsky and the "mature mind" from his letters to Serafima Karchevskaia in ARAN.

"During the fast": Maria Petrova's memoirs, ARAN 767.3.3.

"I remember vividly": Iu. P. Frolov memoirs, Riazan Memorial Museum–Home of academician I. P. Pavlov.

"The more I read": ARAN 259.2.1300/1.

"We are not physicists": PSS, VI, 108.

Chapter 3: The haunted factory

The basic sources for Pavlov's digestive research are his scientific publications, the doctoral theses of his coworkers, and memoirs.

"The digestive canal is in its task": PSS, II, 1, 250.

"What is the activity of this factory": PSS, II, 1, 250.

"When in the mornings": A. F. Samoilov, "Obshchaia kharakteristika issledovatel'skogo oblika I. P. Pavlova," in *I. P. Pavlov v vospominaniiakh sovremennikov*, ed. E. M. Kreps (Leningrad: Nauka, 1967), 203–4.

"One single man's intellectual property": for the Nobel Prize deliberations, see:

Daniel P. Todes, *Ivan Pavlov: A Russian Life in Science* (New York: Oxford University Press, 2014), 252–65; citation on 255.

"The psychic moment" and "scientific flesh and blood": PSS, II, 2, 104.

"Of course, not all experiments are so similar": PSS, II, 2, 42.

Chapter 4: Pavlov's quest

The basic sources for Pavlov's research on conditional reflexes are his scientific publications, the records of his Wednesday meetings with coworkers, his lab notebooks, manuscripts and correspondence in ARAN, and memoirs.

"Man...driven by some dark forces": Olga Yokoyama, trans., *Pavlov on the Conditional Reflex: Papers, 1903–1936* (New York: Oxford University Press, 2022).

"A Napoleonic type": ARAN 259.1.59/4.

"We consider all so-called psychic activity": ARAN 259.1.59/2.

"We elicit various conditions": PS, II, 95.

Chapter 5: Come the Bolsheviks

The material on Pavlov and the Bolsheviks comes from ARAN, the archives of the Communist Party and various state ministries, and memoirs.

"Where are you, freedom": ARAN 259.1a.2.

"We have overextended": ARAN 259.1a.4.

"I was, am, and will remain": ARAN 259.1a.12.

"Oh noble and stern apparition!": ARAN 259.1.108.

"I have complained many times": ARAN 259.1 112; author's interview
   with V. I. Fadeeva, taped interview with N. N. Traugot in ARAN
   razriad XVI.
"We have lived and are living": cited from Molotov's copy of the letter
   in "'Poshchadite zhe rodinu i nas': Protesty akademika I. P. Pavlova
   protiv bol'shevistskikh nasilii," *Istochnik: Vestnik Arkhiva
   Prezidenta Rossiiskoi Federatsii* 1 (1995):138–44, citation
   on 139–40.

Chapter 6: Nervous types

This chapter is based on Pavlov's lab notebooks, his scientific work in
   PSS, the publications of his coworkers, the discussions in PS, and
   memoirs.
"In one of his early works Freud": PS, I, 112.
"an unrestrained choleric": Petrova's memoirs, ARAN 767.3.3.
"cycloid" and "cyclically unbalanced strong type": PS, II, 533.
"Strong, balanced, labile": ARAN 259.1.43.

Chapter 7: Year of climaxes

Material is from ARAN (Pavlov's remarks at Congress, unpublished
   notebooks and manuscripts), memoirs, and Soviet newspapers.
"Looking like a medieval saint" and "came through the earphones":
   John Fulton, *The Trip to Russia, Sweden and Finland*,
   Manuscripts and Rare Books section, GC/71/15, Science Library,
   University College, London; and Cornelia Cannon, *The Friendship
   of Dr. Pavlov and Dr. Cannon*, Walter B. Cannon papers, box 39,
   file 508, Francis A. Countway Library of Medicine, Manuscript
   Division, Harvard University Medical School, Boston.
"You have heard and seen....As you know, I am an experimenter":
   Izvestiia, August 20, 1935.
"Aside from the fact....Why are they all included": cited from
   Molotov's copy in "'Poshchadite zhe rodinu i nas,'" 143–44.
"the elementary associations, or knowledge" and list that follows:
   ARAN 259.1.54.

Chapter 8: Final reflections

"Psychology as a Science": manuscript in ARAN 259.1.66.

"I must admit that the longer your regime exists": Pavlov's various drafts of his essay and letter to Molotov are held in ARAN 259.1a.39.

"What is the most difficult": ARAN 259.1a.39.

"However much I were to conduct myself": ARAN 259.1a.39.

"In order that an omnipotent God protect you": ARAN 259.1a.39.

"You are the successors to Jesus's mission": ARAN 259.1a.39.

"Jesus Christ was a man": Taped interview with N. N. Demin, who reads from a document with this passage from the original stenogram of the meeting, ARAN *razriad* XVI. Pavlov's words were confirmed by others present. This passage was deleted from the published version of PS.

"At the basis of Bolshevism": A. L. Chizhevskii, "O poseshchenii I. P. Pavlova v 1926 godu," In *I. P. Pavlov: Pro et contra*, ed. Iu. P. Golikov and K. A. Lange (St. Petersburg: Russkii Khristianskii Gumanitarnyi Institut, 1999), 463–72; citation on 467–8. Chizhevskii visited Pavlov several times; although he dated this conversation from 1926, internal evidence makes clear that it occurred after the 1935 International Congress.

Further reading

Works about Pavlov's life

Babkin, B. P. *Pavlov: A Biography*. Chicago: University of Chicago Press, 1949.

Eckstein, Gustav. "Ivan Petrovich Pavlov." *The New Yorker*, December 2, 1950, 158–70.

Gantt, W. Horsley. "Ivan P. Pavlov: A Biographical Sketch," in *Lectures on Conditioned Reflexes: Twenty-Five Years of Objective Study of the Higher Nervous Activity (Behaviour) of Animals*, by I. P. Pavlov. Translated by W. Horsley Gantt, 11–31. New York: International Publishers, 1928.

Kellogg, John Harvey. "A Visit to Pavlov's Laboratory." *The Bulletin of the Battle Creek Sanitarium and Hospital Clinic* 24 (1929): 203–21.

Rose, Kenneth, Erwin Levold, and Lee Hiltzik. "Ivan Pavlov on Communist Dogmatism and the Autonomy of Science in the Soviet Union in the Early 1920s." *Minerva* 19 (1991): 463–75.

Todes, Daniel P. *Ivan Pavlov: A Russian Life in Science*. New York: Oxford University Press, 2014.

Pavlov's scientific works in English

Readers in English will soon, finally, have an accurate, thoughtful, excellent translation in Pavlov's lively voice of his articles and talks on conditional reflexes:

Yokoyama, Olga. *Pavlov on the Conditional Reflex: Papers, 1903–1936*. New York: Oxford University Press, 2022.

The two previous standard translations are inaccurate, incomplete, misleading, and wooden:

Pavlov, I. P. *Lectures on Conditioned Reflexes: Twenty-Five Years of Objective Study of the Higher Nervous Activity (Behaviour) of Animals*. Translated by W. Horsley Gantt. New York: International Publishers, 1928.

Pavlov, I. P. *Conditioned Reflexes: An Investigation of the Physiological Activity of the Cerebral Cortex*. Edited and translated by G. V. Anrep. New York: Dover, 1960.

The translation of Pavlov's monograph on digestion is better:

Pavlov, I. P. *Lectures on the Work of the Main Digestive Glands*. Translated by W. H. Thompson. London: Charles Griffin, 1902.

Perspectives on Pavlov's scientific work

Adams, Matthew. "Why Pavlov's Dogs Still Matter," in *Anthropocene Psychology: Being Human in a More-Than-Human World*, 13–47. London: Routledge, 2020.

Fanselow, Michael. "Pavlov's Continuing Impact," in *Pavlov on the Conditional Reflex: Papers, 1903–1936*, edited by Olga Yokoyama. New York: Oxford University Press, 2022.

Gray, Jeffrey. *Ivan Pavlov*. New York: Viking Press, 1980.

Joravsky, David. *Russian Psychology: A Critical History*. Oxford: Basil Blackwell, 1989.

Kimmel, H. D. "Notes from 'Pavlov's Wednesdays': Gestalt Relationships as Conditional Stimuli." *American Journal of Psychology* 89, no. 4 (December 1976): 745–49.

Raikhel, Eugene. "Reflex/Refleks." *Somatosphere*, February 11, 2014.

Rescorla, Robert. "Pavlovian Conditioning: It's Not What You Think It Is." *American Psychologist* 45, no. 3 (March 1988): 151–60.

Ruiz, Gabriel, Natividad Sánchez, and Luis Gonzalo De la Casa. "Pavlov in America: A Heterodox Approach to the Study of His Influence." *Spanish Journal of Psychology* 6, no. 2 (2003): 99–111.

Todes, Daniel P. "From Lone Investigator to Laboratory Chief: Ivan Pavlov's Research Notebooks as a Reflection of His Managerial and Interpretive Style." In *Reworking the Bench: Research Notebooks in the History of Science*, edited by Frederick L. Holmes, Jürgen Renn, and Hans Jorg Rheinberger, 203–20. Dordrecht: Kluwer, 2003.

Todes, Daniel P. *Pavlov's Physiology Factory: Experiment, Interpretation, Laboratory Enterprise*. Baltimore: Johns Hopkins University Press, 2002.

Watson, J. B. "The Place of the Conditioned-Reflex in Psychology." *Psychological Review* 23 (1916): 89–116.

Wells, H. G. "Mr. Wells Appraises Mr. Shaw. He Contrasts the Contribution of the Playwright with That of Pavloff, Russian Scientist, and Asks: To Whom Does the Future Belong: The Man of Science or the Expressive Man?" *New York Times Magazine*, November 18, 1927.

Windholz, George. "Pavlov vs. Koehler: Pavlov's Little-Known Primate Research." *Pavlovian Journal of Biological Science* 19 (1984): 23–31.

Windholz, George, and Wanda Wyrwicka. "Pavlov's Position toward Konorski and Miller's Distinction between Pavlovian and Motor Conditioning Paradigms." *Integrative Physiological & Behavioral Science* 31, no. 4 (1996): 338–50.

Yerkes, Robert M., and Sergius Morgulis. "The Method of Pawlow in Animal Psychology." *Psychological Bulletin* 6, no. 8 (August 15, 1909): 257–73.

Pavlov's context: Russia and Russian science

Cohen, Stephen F. *Bukharin and the Bolshevik Revolution.* Oxford: Oxford University Press, 1980.

Dostoevsky, Fyodor. *The Brothers Karamazov.* Translated by Constance Garnett. New York: Barnes & Noble, 2004.

Fitzpatrick, Sheila. *Everyday Stalinism: Ordinary Life in Extraordinary Times: Soviet Russia in the 1930s.* New York: Oxford University Press, 2000.

Freeze, Gregory. *The Parish Clergy in Nineteenth-Century Russia: Crisis, Reform, Counter-Reform.* Princeton, NJ: Princeton University Press, 1983.

Graham, Loren. *Science, Philosophy and Behavior in the Soviet Union.* New York: Columbia University Press, 1987.

Kozulin, Alex. *Psychology in Utopia: Toward a Social History of Soviet Psychology.* Cambridge, MA: MIT Press, 1984.

Krementsov, Nikolai. "Big Revolution, Little Revolution: Science and Politics in Bolshevik Russia." *Social Research* 73, no. 4 (2006): 1173–204.

Krementsov, Nikolai. *Stalinist Science.* Princeton, NJ: Princeton University Press, 1997.

Krementsov, Nikolai. *With and without Galton: Vasilii Florinskii and the Fate of Eugenics in Russia.* Cambridge, UK: Open Book Publishers, 2018.

Lincoln, W. Bruce. *Passage through Armageddon: The Russians in War & Revolution, 1914–1918*. New York: Simon & Schuster, 1986.

Lincoln, W. Bruce. *Red Victory: A History of the Russian Civil War*. New York: Simon & Schuster, 1989.

Schlogel, Karl. *Moscow 1937*. Translated by Rodney Livingstone. New York: Polity Press, 2012.

Sirotkina, Irina. "When Did Scientific Psychology Begin in Russia?" *Physis; rivista internazionale di storia della scienza* 43 (2006): 239–71.

Todes, Daniel P., and Nikolai Krementsov. "Dialectical Materialism and Soviet Science in the 1920s and 1930s." In *A History of Russian Thought*, edited by William Leatherbarrow and Derek Offord, 340–67. Cambridge: Cambridge University Press, 2010.

Tolz, Vera. *Russian Academicians and the Revolution: Combining Professionalism and Politics*. New York: St. Martin's Press, 1997.

Tucker, Robert. *Stalin in Power: The Revolution from Above, 1928–1941*. New York: W. W. Norton, 1990.

Vucinich, Alexander. *Science in Russian Culture, 1861–1917*. Stanford, CA: Stanford University Press, 1970.

Pavlov's context: scientific traditions

Ash, Mitchell. *Gestalt Psychology in German Culture, 1890–1967*. Cambridge: Cambridge University Press, 1995.

Bernard, Claude. *An Introduction to the Study of Experimental Medicine*. Translated by Henry Copley Greene. New York: Dover, 1957. First published 1865 by Balliere (France).

Boakes, Robert. *From Darwin to Behaviorism: Psychology and the Minds of Animals*. Cambridge: Cambridge University Press, 1984.

Canguilhem, George. *The Normal and the Pathological*. Translated by Carolyn R. Fawcett. New York: Zone Books, 1991. First published 1966 by Presses universitaires de France (Paris).

Coleman, William. "The Cognitive Basis of the Discipline: Claude Bernard on Physiology." *Isis* 76 (1985): 49–70.

Comfort, Nathaniel. *The Science of Human Perfection: How Genes Became the Heart of American Medicine*. New Haven, CT: Yale University Press, 2013.

Davenport, Horace. *A History of Gastric Secretion and Digestion*. New York: Oxford University Press, 1992.

Fearing, Franklin. *Reflex Action*. Cambridge, MA: MIT Press, 1970.

Harrington, Anne. *Reenchanted Science: Science in German Culture from Wilhelm II to Hitler*. Princeton, NJ: Princeton University Press, 1999.

Kevles, Daniel. *In the Name of Eugenics: Genetics and the Uses of Human Heredity*. Cambridge, MA: Harvard University Press, 1998.

Koehler, Wolfgang. *The Mentality of Apes*. Translated by Ella Winter. New York: Harcourt Brace, 1925.

Koehler, Wolfgang. *Gestalt Psychology*. New York: Liveright, 1929.

Smith, Roger. *Inhibition: History and Meaning in the Sciences of Mind and Brain*. Berkeley: University of California Press, 1992.

Museums

For visitors to St. Petersburg, I strongly recommend a visit to the Memorial'nyi Muzei–Kvartira akademika I. P. Pavlova (Memorial Museum–Apartment of academician I. P. Pavlov), located in the well-preserved apartment in which he lived from 1918 to 1936. Pavlov's art collection, study, photos, furniture, and so forth—and the museum's expert guide—convey a palpable sense of the man. Equally worthwhile is the Memorial'nyi Muzei–Usad'ba akademika I. P. Pavlova (Memorial Museum–Home) in Pavlov's boyhood home in Riazan.

Index